T0075419

101 Curious Tales
of East African Birds

101 Curious Tales of East African Birds

A Brief Introduction to Tropical Ornithology

Colin Beale

PELAGIC PUBLISHING

Published by Pelagic Publishing
20–22 Wenlock Road
London N1 7GU, UK

www.pelagicpublishing.com

Copyright © Colin Beale 2023
Photographs © individual photographers as credited on p. 215.

The moral rights of the author have been asserted by him in accordance with the Copyright, Designs and Patents Act 1988.

All rights reserved. Apart from short excerpts for use in research or for reviews, no part of this document may be printed or reproduced, stored in a retrieval system, or transmitted in any form or by any means, electronic, mechanical, photocopying, recording, now known or hereafter invented or otherwise without prior permission from the publisher.

A CIP record for this book is available from the British Library

ISBN 978-1-78427-291-3 Paperback
ISBN 978-1-78427-292-0 ePub
ISBN 978-1-78427-293-7 ePDF

https://doi.org/10.53061/EHMT6812

British Library Cataloguing in Publication Data
A catalogue record for this book is available from the British Library

Layout and typesetting by Trevor Johnson

Printed in the Czech Republic by Finidr

MIX
Paper | Supporting
responsible forestry
FSC® C014138
FSC
www.fsc.org

Cover image: Little Bee-eater *Merops pusillus*. Guy Edwardes/naturepl.com

CONTENTS

INTRODUCTION

This is not a typical bird book. It is more like the random stream of consciousness you get if you sit me down, hand me a cold beer and ask me about the birds I've learned to love in East Africa. There's no start, and no real end – just brief illustrated accounts of 101 birds that might be encountered in a few days of birding in the region, loosely linked by the fascinating stories of nature they have come to epitomise in my mind.

But although there's no particular order to the species and topics covered, that doesn't mean there's no understanding: I'm a scientist both by trade and at heart, so each tale represents in some way the latest scientific understanding of the topic covered (I've provided footnotes linking to the original research articles for those who want to know all the facts, and a comprehensive index of species and topics at the end). Having said that, this book, like the birds themselves, can be appreciated in whatever way you want: enjoy the photographs, flick through the stories, ponder the science – or do them all at once.

Starting life as a series of tweets linked by the #BirdsAtTea hashtag, many of the stories here were originally compiled as a distraction during the UK's first Covid-19 lockdown in 2020: the need for mental escape prompted me to sort through some of the photographs I've accumulated through my research in Tanzania. I hope you enjoy reading these snippets as much as I enjoyed the process of writing them, recalling memories of all the hours in the field that lie behind it all.

LILAC-BREASTED ROLLER
Coracias caudatus

ALL TROPICAL BIRDS ARE COLOURFUL, RIGHT?

Often, the first bird noticed by visitors to East Africa is the Lilac-breasted Roller. Brightly coloured, reasonably common and always sitting on a prominent perch, it seems to confirm everyone's preconception that tropical birds are colourful. Indeed, this particular species looks as though it has been coloured in by a child with a brand new pack of felt-tip pens and no concept of the adage 'less is more'. Certainly, many tropical birds, like the rollers, are brightly coloured, but the reality is that there are plenty of colourful birds elsewhere (the Blue Tits in my garden in York, for instance, are gorgeous) and also many drab birds in the tropics. Until very recent advances in colour analysis, it was extremely hard to test whether birds in the tropics were more colourful on average, or simply that there were more species in the tropics – both colourful and plain – and the proportion of species that were colourful were not that different. Only in 2022 has the matter been laid to rest, for songbirds at least, with the discovery that the average variety of colours shown by a bird in the tropics is about 20–30% greater than for songbirds in the temperate zone.[1] The widest variety of colourful birds tend to be found in the places with the most species and also in darker forests, implying that the reason for tropical colourfulness is the need to distinguish themselves more readily among the masses. Life is genuinely more vibrant in the tropics.

BEAUTIFUL SUNBIRD
Cinnyris pulchella

TROPICAL AVIAN DIVERSITY

Another ridiculously colourful species, the Beautiful Sunbird is emblematic of the extraordinary diversity of the tropics. Nectar is a pretty universal resource provided by plants to encourage the visits of birds such as sunbirds, as well as insects and other pollinators. Ecological theory suggests that unless different species have developed specialisations which enable them to partition a resource like flowers with nectar into distinctive sets (say, flowers with long trumpet shapes, as opposed to flowers with open petals), then one pollinator species should be best adapted to use it, and should outcompete others. However, it doesn't take long watching a Beautiful Sunbird at a patch of flowers to realise that several other sunbird species are also using it – a form of diversity that ecologists find hard to explain. In fact, the extreme diversity of the tropics (where sites of similar area can contain four to ten times as many bird species as in temperate sites) remains a puzzle, with many theories proposed to explain it. Our best current theory links the long-term stability of climate in the tropics (compared to temperate areas) to lower extinction rates,[2] but how quite so many sunbird species coexist with minimal apparent specialisation remains a mystery.

SHELLEY'S GREENBUL
Arizelocichla masukuensis

DIVERSITY OF MONTANE FORESTS

In the forests where it is found, Shelley's Greenbul is a common resident or altitudinal migrant, moving up and down as the seasons warm and cool. Yet it only occurs in a relatively few montane (mountainous) forest patches in East Africa. Those forest patches where it lives are some of the most biologically rich locations of the region, but there are plenty of apparently suitable montane forests where doesn't occur. So what determines the richness of forest fragments and the distribution of Shelley's Greenbul? In most of the world, smaller forest patches hold fewer species, but this isn't the case in Tanzania's Eastern Arc Mountains. Indeed, the tiny forests of East Usambara and the Uluguru ranges have the richest biodiversity. These forests are, however, both found on the geologically oldest hills and also those closest to the sea, leading to a climatic moderation effect. Consequently, like the tropics in miniature, diversity on tropical mountains seems driven by the joint effects of geological age, which allows lots of time for new species to evolve, and climatic stability,[3] which lowers the local extinction rate.

BROAD-RINGED WHITE-EYE
Zosterops eurycricotus

SKY ISLAND SPECIATION

When you roam through forests in the northern mountains of Tanzania, one of the species you keep on bumping into is the Broad-ringed White-eye. These sweet-looking birds illustrate one of the most surprising discoveries about the East African avifauna of recent years: the speed and consistency of evolution in montane forests. We are aware that older mountains harbour more diverse bird communities, partly because such 'sky islands' have had longer to accumulate new species, but until recent advances in genetics we didn't know how this really worked: do montane forest species colonise from other isolated forests, fragment to fragment, and then gradually drift apart, or do lowland species form new highland sisters? By studying the African white-eyes,[4] we have come to learn that most montane white-eye species are not cousins of each other, but rather of the smaller, yellower and widespread lowland species that have repeatedly pushed upwards to colonise the montane forests, each time evolving a darker greenish plumage and broad eye-ring. This evolution of similar features in genetically separate populations is a great example of convergent evolution, but it also means that the similar-looking populations on different mountains are actually very different species that we had been overlooking for many years.

RING-NECKED DOVE
Streptopelia capicola

CONVERGENT EVOLUTION

Probably my favourite example of convergent evolution – the phenomenon whereby different species evolve similar traits completely independently of each other – can be illustrated by the widespread and common Ring-necked Dove. Most of the world's baby birds are fed on invertebrates of one form or another, but young doves and pigeons get milk. Unlike mammal milk, which is simply modified sweat, pigeon milk consists of specially grown cells which are shed from the adult's crop.[5] Despite very different origins, both pigeon and mammalian milk consists of about 60% protein and 40% fat, and both contain antibodies that provide immediate protection against disease,[6] with pigeon milk also stimulating the immune system of chicks. A further similarity is that both milks provide the young with beneficial bacteria that populate the gut and offer assistance to later digestion. That evolution has generated such similar baby food independently through two completely different pathways is a wonderful example of convergent evolution. Once grown up, Ring-necked Doves supply the tireless backing vocals to the soundtrack of the African bush, giving a constant reminder of the inventiveness of evolution.

CRESTED FRANCOLIN

Ortygornis sephaena

MORNING SOUNDS

While the Ring-necked Dove may be the soundtrack of the warmer parts of the day, the Crested Francolin is the alarm clock of the savannah. Groups or pairs of this species are often the first bird to get noisy in the morning. During the breeding season, pairs form bonds through duetting – it's a crazy sound,[7] and all about breeding. But coordinating these pairings requires females to choose her mate from a male flock.[8] Outside of the breeding season, males form coalitions. Females seeking a mate join a male flock and start calling. This gets the males tremendously excited: they all start calling to try and follow her lead or prevent others from matching her song, fights break out, and on it goes. So during the dry season, the big noisy flocks of Crested Francolins we hear are essentially males trying to jam the signal of a female who wants to find her perfect singing and nesting partner. It also probably signals to other coalitions nearby that the territory is occupied. Hence, the morning wake-up calls of Crested Francolins are about both territory defence and finding mates: two of the most common reasons for birdsong. Even if it is more of a cacophony than something nice and melodious.

SPOTTED MORNING THRUSH
Cichladusa guttata

WHY DO BIRDS SING AT DAWN?

A much more melodic dawn chorister is the Spotted Morning Thrush. Like most birds, the main purposes of its song are to attract a potential mate and to signal that the territory is occupied, but why should such messages be best sent out at dawn? There are many theories to explain this: maybe cool air helps song carry? Or perhaps singing in low light allows females looking for sneaky matings with a neighbour to find them easily without being detected? The best theory is that birds often have spare energy left over from overnight reserves, which is best spent singing. Small birds can burn 10% of their mass keeping warm overnight.[9] When temperatures change a lot between nights, birds must have enough fat to survive the coldest night possible, but mostly it isn't that cold and consequently most mornings birds wake up with spare fat. Such excess weight could make them more vulnerable to predators, so they use spare energy singing,[10] neatly explaining the dawn chorus. This also explains why tropical dawn choruses aren't as impressive as those in temperate zones, because temperatures don't vary so much night to night, allowing birds to plan their weight gain more accurately.

RED-AND-YELLOW BARBET
Trachyphonus erythrocephalus

DUETTING

As well as having amazing plumage, the termite-loving Red-and-Yellow Barbet illustrates one of the surprisingly common features of African birdsong: the relatively high frequency of duets. Actually, this species is very kind to struggling ornithologists as pairs duet loudly, calling something that almost sounds like 'red and yellow, red and yellow'.[11] Bird duets come in two types, a tightly connected call and response that can almost sound like a single bird but is usually a pair, or, as with the Red-and-Yellow Barbet, simply two or more individuals simultaneously shouting their heads off (or, as described in a paper on this subject, 'two autochthonous periodicities which are only in part responsive'[12]). So why should duetting be more common in African bird species than in temperate ones? We're guessing a bit, but duets seem to be important for cementing the pair bond, and pair bonds appear to be stronger in tropical species because individuals tend to live longer than northern temperate species, making investment in pair bonds more valuable than in shorter-lived birds that are likely to only be together for a year or two at most.

BAR-THROATED APALIS
Apalis thoracica

TROPICAL BIRD SURVIVAL

The Bar-throated Apalis is a rather cute member of the cisticola family, found in forest edges across eastern Africa as far as South Africa, and it is thanks to studies of birds like these that we know that tropical species tend to live longer than equivalently sized northern temperate species. Each year, just over two thirds of adult Bar-throated Apalis living in tropical East Africa survive to the following year.[13] By contrast, for a similar-sized northern temperate species such as the Blue Tit, just over half of adults survive to the next year, or of a migrant like the Lesser Whitethroat, not quite one third survive. These differences sound small, but mean tropical species live an average of twice as long as their northern temperate counterparts. Until recently, ornithologists spent a great deal of time trying to explain this apparently general pattern.[14] Now, however, we know that in temperate zones of southern Africa and of South America birds actually live as long or longer still than their tropical counterparts. Even the same species live longer in the southern temperate zone: three quarters of adult Bar-throated Apalis survive each year in South Africa.[15] This demonstrates how a northern bias in research can lead scientists to ask entirely the wrong questions: we should be considering why birds in on southern continents live longer than northern ones, not simply tropical species.

WIRE-TAILED SWALLOW
Hirundo smithii

OPTIMAL CLUTCH SIZE

While information from southern temperate zones shows us it is a myth that tropical birds generally live longer than temperate birds, the Wire-tailed Swallow is the species that taught us that tropical species really do lay fewer eggs than their temperate counterparts. Nesting of Wire-tailed Swallow was first studied by the colonial-era English ornithologist Reginald E. Moreau at Amani in Tanzania. He noticed that many birds in Tanzania laid fewer eggs than near relatives in temperate zones.[16] Wire-tailed Swallows lay two to three eggs but, for example, Barn Swallows lay four to five. Meanwhile, in the UK one of the forerunners of modern ornithology, David Lack, had shown that birds lay as many eggs as they can raise young. If he added an egg to a nest, the parents may feed the chick to fledging, but more chicks died after fledging so the enlarged clutch resulted in fewer young. Why then, would African birds raise fewer young? The first theories were that tropical birds might restrict their breeding to prevent overpopulation, but evolution makes birds do what is best for themselves, not what is best for the population. Instead, and despite ideas of tropical abundance, the smaller clutches of tropical species probably reflect either lower food supply in the tropics or higher survival of juveniles in less seasonal environments.[17]

WHITE-FRONTED BEE-EATER
Merops bullockoides

SOCIAL NESTING

Another difference between tropical and temperate birds is the frequency of cooperative breeding in the tropics. Few species illustrate this behaviour as well as the White-fronted Bee-eater, which was one of the first cooperative species to be studied.[18] These birds nest in colonies of around 200 pairs. Within the colony, related males nest close to one another, forming a 'clan', and among the clan they may share up to seven related 'helpers' – usually offspring from previous nests – that assist in feeding chicks, rather than raising their own young. Pairs defend their preferred foraging territories away from the colony and will tolerate other clan members within it, but they fight off visitors from other clans, creating an exclusive clan foraging territory. However, this is not a totally happy family: as pairs commence nesting, nearby relatives that already have nests will hassle the new pair, trying to encourage the new pair to give up and help them raise their own chicks instead. Such harassment is a common cause of nest failure. As with most birds, females disperse between clans (and colonies) to find a mate, so males within a clan are more closely related than females, which has an important impact on what happens if a pair's nesting attempt fails. Males, being closely related to other clan males, share lots of genes with the chicks of their neighbours: if they can't raise their own young, helping their neighbours is favoured by evolution. Females, however, aren't closely related to others in the clan, so may instead parasitise their neighbours by laying eggs in another nest. This occurs in around one in six nests,[19] showing quite how complex avian social lives can be.

VULTURINE GUINEAFOWL
Acryllium vulturinum

SOCIAL STRUCTURE

Complex social lives are often associated with high intelligence, but the social life of the (not desperately clever) Vulturine Guineafowl shows how exceptions exist. This species is common in the drier corners of East Africa, where it lives in flocks of 30 to 60 birds. Despite their small brains, these birds have multilevel social systems similar in complexity to those of elephants.[20] The first level of Vulturine Guineafowl society is the family group: male and female plus offspring. Each family joins with other families to form the flock that is usually found together. And during the wet season when resources are plentiful, flocks sometimes connect with other flocks too: but always the same ones, just like the clan structure of elephants.[21] All three levels of family, flock and super-flock are stable over time and are largely peaceful. Stable, three-level social hierarchies in elephants have been thought to require substantial intelligence for individuals to recognise and remember their more distant friends, but their existence in the very small-brained Vulturine Guineafowl suggests that maintaining multi-level societies must be simpler than realised. Such complex social structures contrast with most animals' family group structures, or more casual associations. But complex structure may enable cultures to form and help altruism evolve.[22] Certainly, if you watch them in their flocks you'll see that these birds help each other more than most species do.

SOUTHERN GROUND HORNBILL
Bucorvus leadbeateri

FEMALE DISPERSAL

If there are any other birds to compare to elephants, it might be the Southern Ground Hornbill for large size, sociability and longevity. These birds stand around 1m tall and can live up to 70 years, breeding in social groups of three to seven individuals. Each family group requires a large tree hole to nest in, and suitable sites are always in short supply. It also takes a very long time to raise young: a pair only lays eggs every two or three years and then the young need two years of feeding.[23] Females disperse to another group to seek a mate from three years of age, while males may stay with their parents until as much as nine years old (and eventually inherit the nest hole). This illustrates a general difference between birds and mammals: most social mammals, such as elephants, form stable female groups from which males disperse, while birds generally do the opposite.[24] Though curiously, if female Southern Ground Hornbills do stay with their parents to help raise young, they give more assistance than their brothers. Why there is a difference in the sex that disperses between mammals and birds isn't clear. The main theory is that in general male birds value territories more than they do mates, while in mammals the opposite is true,[25] but I'm not sure that does much more than raise a further question of why such a difference exists in the first place.

RATTLING CISTICOLA
Cisticola chiniana

FLEXIBLE BREEDING SYSTEMS

The decisions birds make about when and how to breed are rarely more clearly defined than in the case of the Rattling Cisticola. This is one of the commonest species in bushy areas, where noisy groups are widespread. The population of Rattling Cisticolas varies year on year: after a few years with good rains they are common, but following a drought numbers may be much reduced.[26] At low densities, their territorial behaviour is fairly standard: each male will defend the largest territory he can, and it may contain one or two females who nest within his patch. After a few good, wet rainy seasons, however, the population grows and males find themselves spending more and more time defending their territory from ever more neighbours. At some point defence simply becomes impossible, so he gives up and makes an alliance with one or two unrelated males to defend a joint territory together. In this combined territory four or five females may nest, each male fathering chicks in each nest. Such communal territoriality is unusual, but shows how the costs and benefits of defending territories are often dependent on environmental conditions. These may not be the prettiest birds around but they've got character, and social interactions among communal groups are fun to watch!

SUPERB STARLING
Lamprotornis superbus

SEX RATIO ALLOCATION

The Superb Starling illustrates how birds have evolved remarkable strategies to deal with variable environmental conditions. Most animals maintain a one-to-one ratio of males to females. While even sex ratios overall are the evolutionarily stable position, some birds can adjust the sex ratio of their chicks to their individual advantage. For example, if the eventual potential breeding success of chicks relates to parental or environmental conditions at birth, active adjustment of sex ratios would be favoured by evolution. Unlike the case with mammals, in birds, females determine the sex of eggs, and can bias determination one way or the other.[27] If one sex has more variable breeding success than the other – for instance, if one dominant male prevents other males from mating, he will have very high mating success and the others very low, so variation in mating success is high among males; in the same species, though, all females may succeed in raising chicks each year with the same male, so there is lower variation in breeding success of females than males – then it makes sense to invest in babies of the sex with highest variation only when conditions are optimal for producing the biggest and strongest offspring: those most likely to dominate breeding opportunities. Male variation is often greatest, but in cooperative breeders such as Superb Starlings, variation is greatest in females, hence in wet years with much food, 70% of chicks are female,[28] and vice versa in dry years. So, Superb Starlings not only look great but show how evolution works to fine-tune sex ratios based on environmental condition. This ensures a one-to-one ratio overall, but maximises the chances of producing high-quality individuals of the sex where quality matters most.

RED-CHESTED CUCKOO
Cuculus solitarius

SEX DETERMINATION IN BIRDS

Female birds can bias sex ratios because in birds females have two different sex chromosomes, a difference from mammals that also enables remarkable evolution in species such as the Red-chested Cuckoo. In East Africa this bird is the harbinger of rains, as its 'it will rain' call is a common sound at the start of the rainy season. Like most cuckoos, it is a brood parasite, laying eggs in the nests of a variety of robin-chat species. Individual female cuckoos lay only one type of egg, which closely matches the eggs of its own foster species, but there are at least three colours and patterns of egg laid by Red-chested Cuckoos that each match different hosts.[29] Because female birds have unmatched sex chromosomes, one of which comes only from her mother, egg-pattern genes on the uniquely female sex chromosome will only ever pass through the female line. Consequently, a female can mate with any male but still create eggs that match their specific host – even in terms of how they appear in the UV spectrum.[30] Matching eggs well is important:[31] thrushes are the same size as cuckoos, but are great at removing cuckoo eggs and are rarely parasitised, while smaller robin-chats are far less apt in this department and hence are often victim to the trick. There's a theory that the original cuckoo hosts were larger birds, but these species got so good at spotting cuckoo eggs that cuckoos lost the competition. So, the Red-chested Cuckoo reveals how birds have different sex-determination chromosomes from mammals, and while male mammals let the Y chromosome degenerate, female birds have put their odd chromosome to great use, allowing evolution of independent female strains while males enable good gene-flow across the species.

PIN-TAILED WHYDAH

Vidua macroura

RUNAWAY SELECTION

Alongside cuckoos, whydahs are a second important group of brood parasites in East Africa. The Pin-tailed Whydah is the most widespread of the group, and it lays eggs in the nests of a variety of waxbill species. Breeding costs brood parasites virtually nothing: males provide only sperm, and females must grow eggs, but once these have been laid neither sex has any further role in parental care. For a female looking for a mate, therefore, all she cares about is ensuring her offspring have the genes most likely to result in future breeding success. This leads to intense sexual selection, as females assess which male has the best genes. The typical story suggests that this could be assessed by an honest signal: if males can only grow long tails because they are both adept at avoiding predators and good at finding the food resources necessary to grow such a tail, females should prefer the males with the longest tails. In reality, of course, the story is more complicated. Probably because of past selection pressure, there is almost no difference between tail length in males today (although there is a difference in the rate at which tails grow, which probably indicates foraging ability[32]). Instead, these long tails are probably maintained as a consequence of genes for both female preference and male tail length becoming physically linked and causing positive feedback,[33] leaving no remaining quality signal at all – a process called Fisherian runaway evolution.

GREATER HONEYGUIDE
Indicator indicator

HUMAN MUTUALISM

The last group of major brood parasites in East Africa are the honeyguides, which lay eggs in barbet and bee-eater nests (honeyguide chicks slaughter their foster siblings with an evil-looking hooked beak), though the Greater Honeyguide is best known for its mutualistic behaviour with humans. Greater Honeyguides often approach people on bush walks. They give a rattling call, flash their white outer tail feathers and fly to the next bush. In some places, people use a specific whistle to call the birds, and then follow them to a beehive. In Mozambique, when human honey gatherers call honeyguides, they double their chances of being guided by birds, and more than double their chances of finding honey.[34] Most people that follow honeyguides reward the birds with the wax or empty honeycomb, a rare example of a human–bird mutualism. In Tanzania, however, the Hadza call and follow, but then bury or burn remaining honeycomb leaving the birds to go hungry so they will continue guiding![35] Thus, the Greater Honeyguide shows us a rare example of a human–bird mutualism, but also how mutualisms can easily be manipulated by one partner. Since these birds evolved over three million years ago, the mutualism probably originally arose with our early pre-human ancestors, but remarkably it has continued through a succession of hominins to the present day.

RED-BILLED OXPECKER
Buphagus erythrorynchus

MORE MUTUALISM

Probably the best-known mutualism between bird and mammal is that of the oxpeckers and their mammalian hosts. With their Yellow-billed cousins, Red-billed Oxpeckers are commonly seen wherever large mammals occur. Oxpeckers love eating ticks (an adult bird consumes around 100 adult ticks or up to 1,000 nymphs per day[36]) and ticks are really bad for mammals, spreading disease and slowing calf growth (for instance, delaying growth in young impala by up to 44kg/yr[37]). Thus, it should seem obvious that since the Red-billed Oxpecker eats ticks and in the process the mammals have their ticks removed, both are benefiting: the very definition of a mutualism. Oxpeckers also warn their hosts of potential danger with a distinctive alarm call when they see predators, and giraffes even provide cosy roosts for oxpeckers in their 'armpits'![38] However, just like the story with humans and honeyguides, all is not so simple. Oxpeckers also feed directly on their host's blood, and will keep old wounds open, even enlarging them for regular supplies. While the benefit of tick removal is easily measured (impala with oxpeckers have fewer ticks and need to spend less time grooming), the impact of wound-opening is hard to quantify. If the relationship was completely rosy, we'd expect mammals to welcome oxpeckers, but they often don't.[39] So, while the oxpecker–mammal interaction is largely mutualistic,[40] blood-eating by vampire birds can tip it towards parasitism in wounded mammals. They also consume earwax, droppings, urine, lice, mites, flies, hair and other secretions. Nice!

FORK-TAILED DRONGO
Dicrurus adsimilis

WHEN MUTUALISMS GO WRONG

As well as having a cool name, the Fork-tailed Drongo provides another example of how easily mutualisms can go wrong. These birds are vocal mimics that have turned their talent for mimicry to deception. Drongos spend around a quarter of their time following other birds or mammals, such as the dwarf mongoose, picking up insects and small mammals fleeing these other foragers. As they do, they watch for predators, and have a series of alarm calls for when they spot danger. The animals they are foraging with know the calls and listen out, reducing the amount of time they have to spend looking for predators when drongos are foraging with them. This seems like classic mutualism. If a mongoose should find a particularly tasty snack, however, the drongo will give a warning call, then swoop in and steal the food.[41] If they give false alarms too often, they'd be ignored like the boy who cried wolf, so there's a limit to how often they can deceive their partners. But to increase the frequency with which they can steal without their targets catching on, they use mimicry: when making a deceptive false alarm, they usually mimic an alarm call of the species they're foraging with, which they do less often when sounding a real alarm.[42] Drongos even change the type of alarm they give if they notice that their targets are stopping responding to a given call. This level of deceptiveness implies that they might have a basic 'theory of mind' and are able to interpret what other animals are thinking – strong evidence that at least some bird species have high intelligence.

RED-CAPPED ROBIN-CHAT
Cossypha natalensis

MIMICRY IN SONG

Copying other animals' calls and using them to deceive is extremely unusual, but mimicry and incorporation of other birds' songs is common in species like the Red-capped Robin-chat and friends. Red-capped Robin-chats are seasonal migrants in the forests of Northern Tanzania – and during the breeding season, unexpected calls and songs can often be traced back to this species. Several theories have been proposed to explain why some species incorporate mimicry into their song repertoire. Such behaviour can be broadly classified into three groups: most songbirds have to learn their songs as chicks, so it may simply be accidental copying from nearby birds; alternatively, mimicry may reduce competition with other species; or finally, it may increase male attractiveness to females.[43] With mimicry common in Robin-chats, it seems unlikely that this is an occasional mistake. Competition with others may be reduced if mimicry either convinces other species that a territory is occupied, or sounds sufficiently like a predator or brood parasite to scare others away. If this were happening, we'd expect that the proportion of calls copied from predators or other potential competitors would be more frequent than in the background soundscape, but this doesn't seem to be the case.[44] This implies that such behaviour in Robin-chats has evolved as a sexually selected trait, where mimicry makes males more attractive to females. Although we don't know why this might be the case specifically for Robin-chats, in other mimetic species, females prefer males with more complex mimetic songs because this both demonstrates that the male is clever enough to learn and that he has lived long enough to accumulate a wide repertoire: both indications that he might be good at feeding young, too.

NORTHERN PIED BABBLER

Turdoides hypoleuca

LEARNING AND LANGUAGE

While incorporating mimicry into song might be a basic indicator of intelligence, some birds, such as babblers, may be using vocalisations to communicate in much more profound ways than we might expect. The very noisy Northern Pied Babbler is a common sight in Northern Tanzania, with family groups of up to 20 spending much of their lives patrolling territories. Usually heard before they're seen, these birds are so noisy that they even gain their name from this trait. Thanks to studies on some of their relatives, we are now beginning to understand more of the intelligence shown by these birds. Much of the research into babblers has been on the Southern Pied Babbler; some behaviours may be different from those of the Northern, but a good deal of the vocalisation is probably similar. My two favourite findings from this research are, first, that chicks are actively taught the meaning of calls. Parents bringing food to the nest give a distinctive 'prrrr' call as they approach, and their young learn to associate this call with food. Parents use this call later to help the chicks as they start to forage themselves.[45] Secondly, babblers have an actual language – of sorts.[46] A different call is used when 'discussing' a change in foraging location and combinations of calls have different meanings than simply adding the two together,[47] in the same way that human phonemes make up a word. So all the racket we hear when a family of Northern Pied Babblers comes through probably isn't just babbling: there's actually much more subtle communication taking place.

AFRICAN GREY FLYCATCHER
Melaenornis microrhynchus

TOOL USE AND INTELLIGENCE

Intelligence is often associated with tool use, so the knowledge that the African Grey Flycatcher is one of very few species known to use tools in the wild should make us consider bird intelligence further. Just like Chimpanzees, African Grey Flycatchers have been recorded using pieces of grass to poke in holes and fish for termites.[48] Other birds known to engage in the use of tools include crows, tits and nuthatches, all of which have relatively big brains for their body size. But the African Grey Flycatcher doesn't have a large brain and it is hard to measure intelligence in something so different from us as a bird: if the tool use is completely instinctive, is it any more of an indicator of intelligence than a long beak would be?[49] With the exception of some crow species (and chimpanzees) which use tools and also demonstrate skills in planning and problem solving, the evidence that tool use implies intelligence is mixed. Similarly, there is evidence to suggest that not all intelligent birds use tools, at least not in the wild. So, although the tool use of the African Grey Flycatcher is certainly cool (and worth looking out for) it doesn't necessarily mean they are the brainiest of birds. On the other hand, they seem cheerful types and certainly have an intelligent gleam in their eye, so who knows.

ORANGE-BELLIED PARROT
Poicephalus rufiventris

INTELLIGENCE AND LONGEVITY

When it comes to avian intelligence, parrots such as the African Orange-bellied Parrot are often top of the list. This species is a common resident of baobab-studded drier bush areas of East Africa. As well as high intelligence,[50] parrots also live a very long time for birds. In most animals, size predicts lifespan: small animals die young, big ones live longer. Parrots are an exception: the oldest well-documented parrot was 83 when he died,[51] and lifespans similar to humans are common. Generally, small animals have many predators, larger ones fewer. If an animal is likely to get eaten before too long, there is no point in investing in anti-aging products: it is very likely to die before it can benefit. Hence, smaller animals with many predators tend to spend lots of energy having babies and then die young. Parrots, however, being very clever, can learn to avoid predators better than most birds.[52] With lower predation rates, it makes sense for them to invest in mechanisms that reduce ageing effects, which they have achieved mostly by evolving molecular mechanisms to repair DNA and cellular damage.[53] This affords them remarkably long lives for their body size.

PIED CROW
Corvus albus

WINNERS IN THE ANTHROPOCENE

The intelligence of crows has been known for a very long time: Aesop wrote about the crow's problem-solving behaviour of placing pebbles in a pitcher to raise the water level until it could drink successfully, and it is this intelligence that has allowed the Pied Crow to become one of the great winners of the modern age. Pied Crows are among the relatively few birds that we know were at Olduvai Gorge in Tanzania around 1.3 million years ago. Today, Pied Crows are commonly encountered in cities and towns, and only rarely in pristine bush. The bones found at Olduvai (in the same layers that hold remains of our relatives *Homo erectus* and *Paranthropus boisei*[54]) reveal how it lived a pre-city life. Back then, Pied Crows must have been scarce woodland dwellers, only becoming common when we built towns. Such urban transitions are achieved by intelligent species such as crows,[55] whose cleverness lets them exploit the new opportunities provided by living alongside humans. And crows really are bright: they can solve a problem by looking at it, visualising the tools they need, then creating them, even if the first step is to create a tool that is only useful to help them access another tool that then solves the problem.[56] So, the Pied Crow shows how human environments offer unique opportunities that benefit species capable of exploiting them and reminds us that many of the species common today may once have been much rarer when our human footprint was smaller.

PANGANI LONGCLAW
Macronyx aurantiigula

CLIMATE CHANGE WINNERS

Another species that is expanding in the shadow of human activity is the Pangani Longclaw. This attractive bird is related to the much duller-looking pipits, and to me they epitomise the savannah bird species that are benefiting from anthropogenic climate change. Since around the turn of the Millennium, this species has spread west, crossing the Rift Valley and arriving in Serengeti.[57] As Pangani Longclaws have grown their range they have come into contact with the closely related Yellow-throated Longclaw. Because finding the first record of a species is much easier than being certain that the last individual has vanished, it is much harder to document loss of species than the arrival of a new one, but as Pangani Longclaws moved into south-east Serengeti, Yellow-throated Longclaws seem to have withdrawn towards Lake Victoria. Some of my own research has demonstrated that expanding distributions of savannah birds map onto places where dry-season length has been increasing,[58] but why this matters is unclear as it must be mediated by changes in the habitat that we haven't yet noticed. Our best guess is that maybe altered rainfall patterns alter grass quality, which in turn affect invertebrates.[59] Why Yellow-throated Longclaws should leave as Pangani arrive is more mysterious still. It seems unlikely that the two species differ so precisely in climate preference, so perhaps it is direct competition? Or maybe Pangani is host to a parasite to which Yellow-throated is sensitive? Separating these different processes in order to understand what it is that limits distributions is tricky science. Ultimately, climate change seems to be behind both species distribution shifts, but the Pangani Longclaw illustrates the challenge of providing precise explanations for natural patterns. This is a charming species though, and I propose it as the poster bird for climate change in the savannah.

WHITE-TAILED CRESTED FLYCATCHER

Elminia albonotata

CURRENT CLIMATE CHANGE IN MONTANE FORESTS

Away from the savannahs, climate change impacts are largely adding to the threats to birds, particularly in the upland forests where White-tailed Crested Flycatchers live. This species is a member of the pleasingly named fairy flycatcher family and is regularly heard – but rather rarely seen – in forest understories, particularly at higher elevations. As global temperatures warm, the cooler climates preferred by montane bird species are found at ever-higher elevations; meanwhile, the lower limits of montane birds in East Africa have been retreating up the mountains by an average of 45 metres each decade.[60] Since mountains are generally conical, as birds move up they are focused into smaller and smaller geographical areas, until too much warming could result in no suitable climate for them at all. For species like the White-tailed Crested Flycatcher, we not only know they are retreating up the mountainsides but also that populations are declining largely as a result of reduced recruitment of juvenile birds,[61] itself associated with temperature change. This suggests that climate change impacts in tropical forests may be reducing populations by limiting food availability for juvenile birds. With forest loss on mountains already threatening many species, the added impact of climate change is not good news.

BAGLAFECHT WEAVER
Ploceus baglafecht

PAST CLIMATE CHANGE IN MONTANE FORESTS

Climates have always changed, of course, and the distributions of forest species such as Baglafecht Weavers are testament to the shifts birds can adapt to. This forest-edge species is widespread on the mountains of Africa, from Tanzania to Nigeria. There are eight subspecies of Baglafecht Weaver, each living on its own set of African mountains, mostly well separated by lowlands with no suitable forests. Each population looks different today, but they share common ancestors and must have been connected some time in the past. During ice ages, Africa gets cooler and drier. Around 1.7 million years ago, Africa was very cold and dry, and montane-type forest was much lower and spread across the lowlands of the continent.[62] As temperatures warmed, birds in these forests advanced up the mountainsides, forming isolated populations surrounded by inhospitable lowland. Today, those populations face further climate change, pushing the temperatures higher than they have been for millions of years, and thus plants and animals like Baglafecht Weavers that live on the tops of tropical mountains as refugees from previous cooler climates may have nowhere else to go, making them especially threatened by climate change.[63] Fortunately though, for now Baglafecht Weavers are still common and very attractive birds.

GOLDEN-BACKED WEAVER

Ploceus jacksoni

NESTS AND THE EXTENDED PHENOTYPE

Weavers are usually known for their extraordinary nest-building skills, and the Golden-backed Weaver combines a great nest with rather attractive plumage. Weaver nests are one of the most sophisticated pieces of ornithological architecture and offer a great example of what ecologists call the 'extended phenotype'. Phenotypes are what an organism ends up looking like as a consequence of the interaction between its genes and the environment; therefore, this is what evolutionary selection works on. The idea of the extended phenotype is simply that selection can act to hone external creations made by animals, not just their physical selves. In this case, it applies to tthe nest. And weaver nests are remarkable productions, carefully tied to the tree, usually using half-hitch knots.[64] The process is entirely innate: a weaver raised by hand without seeing another weaver's nest can produce reasonable nests when mature – although if they can watch others they learn the process faster.[65] During a breeding season, males often build many nests and females will visit and destroy inferior examples: one male Southern Masked Weaver created between 22 and 36 nests each year![66] So, females make their choice of mate not just by looking for the most attractive male but also by considering who builds the best nest, ensuring strong selection for nest quality. Male weavers thus build dozens of nests each year, and suffer almost endless disappointment in their attempts.

RED-BILLED BUFFALO WEAVER
Bubalornis niger

SPERM COMPETITION

My favourite weaver story is that of the Red-billed Buffalo Weaver. These are colonial birds that build large, thorny multi-chambered nests in baobabs. Male competition to breed with females is intense. The males fall into one of three types: those with nests, those without nests but resident in the colony tree, and what are known as satellite males – who hang around near the colony. Some nesting males form a coalition with another, so two males might share a nest used by as many as six or seven females. This breeding system means that multiple males often mate with each female, so competition to father chicks isn't just about space in the nesting tree but may continue after copulation. Unusually for birds, males of this species have a false penis that is not used to deliver sperm, but instead to stimulate the female's cloaca during copulation.[67] When a female is ready to mate, she invites her chosen male to a secluded spot away from the colony tree. Instead of the usual fractions of a second for bird copulation, Red-billed Buffalo Weavers will copulate for around 30 minutes, during which the male uses his false penis to rub the female's cloaca until he appears to have an orgasm, when sperm is delivered. To the females, size matters: males with nests have a larger false penis and are generally chosen by females for mating, while satellite males have the smallest organs. But this organ isn't for placing or removing sperm, it's about stimulating the female so she likes the male enough to use his sperm to fertilise her eggs. The fact that females often control mate choice was neglected by – largely male – scientists for far too long.

WHITE-BROWED COUCAL
Centropus superciliosus

SEX ROLES

A rather surprising story of mate choice is found in the White-browed Coucal. Coucals are non-parasitic cuckoos, and are well known for the chick's defence mechanism of vomiting a sticky, black foul-smelling goo on anything that happens to attack the nest. But in some species of coucal, the sex roles are unusual too. In White-browed Coucals, females are bigger than males, and the males do most of the incubation.[68] Females also have lower voices,[69] a trait that seems to be sexually selected such that females compete to sing the lowest song. When they're nesting, if the female is removed, males work twice as hard and the young do fine; remove males, on the other hand, and young don't grow well.[70] This suggests that females may raise more young if they paired with one male, laid eggs for him, then abandoned him to raise them while going off to find another mate. In the Black Coucal this is exactly what happens, but in White-browed the females tend to help out a bit. That females do stay and care for chicks implies that there's a shortage of males: if a female leaves she would be unlikely to find another male to mate with anyway. Other coucal species have populations with more males than females, demonstrating the complex and rapidly evolving nature of avian breeding systems.

GREATER PAINTED-SNIPE
Rostratula benghalensis

REVERSED SEXUAL DIMORPHISM

In some species, such as the Greater Painted-snipe, it is not just the caring roles that are reversed, but the tendency for males to be brighter than females too. This wading bird (not actually a true snipe at all) is one of very few species where the female is more brightly coloured than the male, and where the usual parental roles are completely reversed, such that once the clutch has been completed females never return to the nest, heading off seeking other males to mate with.[71] Care of eggs and chicks only by males is very rare: other than in painted-snipes, I only know of jacanas (close relatives of painted-snipes), Eurasian Dotterel, phalaropes, Black Coucal and button-quails. As is often the case with behavioural ecology, parental care was first studied by male scientists. They concluded that the reasons males usually provide less care than females was too obvious to study in detail: sperm costs little in the way of energy or nutrients to produce, but eggs are a lot more expensive, so if one parent is all it takes to raise young, males should always abandon females (who have already invested more heavily in the nesting attempt) and should head off to look for other partners whenever possible. It took female scientists to point out the fallacy: the costs of initiating breeding are more than just eggs and sperm – territories may need defending, nests building and so on. Moreover, female birds can lay multiple clutches and neither parent will be successful if both were to abandon after laying. Continuing a bad job simply because you have invested lots so far is the sunk-cost fallacy, and female birds are much too clever to fall for this. The Greater Painted-snipe's reversed parental care shows that females abandoning care to males works fine; so, the real question is why it isn't common in other species too? The only hint we have so far is that male-only care is most likely to arise in species where females are the rarer sex.[72]

JACKSON'S WIDOWBIRD
Euplectes jacksoni

TYPICAL SEXUAL DIMORPHISM

Much more typical of species that exhibit single-sex parental care is Jackson's Widowbird. These weaver relatives leave all the care to females, with male contributions limited to sperm. As with the Pin-tailed Whydah, this leads to intense sexual selection – and plumage differences are stark between the jet-black males with their long, curved tails and cryptic brown females. Males facilitate female selection by displaying together in groups at fixed locations called a lek.[73] Each male defends a tiny territory on the lek, and flaunts his virtues with a unique and, to us, rather comic jumping display, performed on specially prepared dance courts. With their long bushy tails puffed up behind them, groups of males bouncing together over their grassland lek are visible to females from far away: the more males that display together, the more females visit. Once females arrive at a lek, the males with the longest tails get the most matings. Unlike in Pin-tailed Whydah, tail length currently seems to be an honest signal of quality, with long-tailed males generally being relatively heavy for their body size. This suggests that in Jackson's Widowbirds females are selecting for good genes, but how long will it be before evolution finds a way to grow long tails without needing to have good genes remains to be seen: such strong selection pressure from females would really benefit a male who could cheat.

KORI BUSTARD
Ardeotis kori

LEKS AND PORPHYRINS

Another group of species known for lekking behaviour are the bustards, of which the Kori Bustard is the largest. At up to 20 kg mass, the Kori Bustard is Africa's heaviest flying bird, and still relatively easy to find in many grasslands of East Africa. During the breeding season, males gather at leks and strut about, fluffing up their white throat and undertail feathers. Compared to the widowbirds, male bustards look rather drab to us, though the white feathers still attract attention from afar. To other bustards, however, the display is much more remarkable: bustards fluoresce pink under ultraviolet light! This colour comes from one of the most common natural pigments: porphyrins. These are what give blood its red colour, but only a few birds (plus hedgehogs and a handful of fish[74]) use it as external pigment. Most birds (but not raptors or mammals) see ultraviolet very well, so bustard displays as viewed by other bustards must be an extraordinary technicolour experience. There may be another trick here too: bustard down feathers have pink porphyrins only for a few minutes before they decay in sunlight. A bustard display that shows pink feather bases allows females to estimate how many displays the male has made: bright pink males are likely virgins and will have high-quality sperm. Females will visit, inspect the displaying males, choose the one they want, mate and then head off to lay eggs. The females are paragons of single motherhood: during incubation they barely leave the nest for 23 days and defend eggs and chicks violently.[75] Discovery of ultraviolet signals in bustard displays is relatively recent, and serves to remind how different birds are from us.

AUGUR BUZZARD
Buteo augur

ULTRAVIOLET VISION IN BIRDS

Once people realised that birds can see ultraviolet, some interesting stories emerged about how raptors such as buzzards and kestrels may use ultraviolet signals to track prey. The argument goes that rodent urine reflects ultraviolet light, which birds can see. Being able to perceive the scent marks that rodents leave to communicate among themselves makes finding rodent prey much easier – and, indeed, rodents make up most of the Augur Buzzard diet.[76] The story begins with an experiment where either rodent urine or water was sprayed onto mouse runs and the effects on birds and rodents monitored.[77] Plots with more urine attracted more buzzards and kestrels and the rodents had shorter lifespans, suggesting the mechanism could work. However, later work looking at raptor eyesight discovered that of all birds, raptors have the worst ultraviolet vision. What's more, rodent urine doesn't really reflect ultraviolet very much, and certainly not in a way that would be useful to a raptor.[78] So what explains the patterns from that first experiment is now unclear. On the other hand, ultraviolet vision in birds other than raptors is very helpful for them: flowers and fruit reflect ultraviolet, and small birds that are otherwise cryptically coloured can signal to each other with ultraviolet patches that make them bright – without, as it turns out, attracting the attention of raptors. It's also nice to be reminded how science eventually corrects itself, although the myth keeps surfacing occasionally.

GABAR GOSHAWK

Micronisus gabar

SPEED OF VISION

While birds of prey may not have ultraviolet vision, species such as the Gabar Goshawk do have exceptional eyesight in other ways. Like other goshawks, the Gabar Goshawk is a pursuit predator that flies at up to 50 km per hour to catch and kill smaller birds in flight. Dashing through dense vegetation at such speeds it is a wonder they don't crash. As well as short wings and a long tail ideal for rapid direction changes, goshawks have two tricks to avoid crashes: they have much higher speeds of visual perception than us and they are happy to fly 'blind'. The speed of visual perception is measured as frequency with which successive images fuse to give an appearance of continuous movement. If we watch an old film, the action seems jerky, because they were filmed at around 18 frames per second. Modern films are faster (at least 24 frames per second) and seem smooth to us, as our eyes perceive at a maximum of 40 frames per second – any movement faster than that is blurred. Birds such as the Gabar Goshawk can see up to 120 frames per second,[79] so high-speed movement is much clearer for them. This better movement perception helps, but if a goshawk shoots past a tree and then discovers a barrier only 5 m ahead it will still crash unless it can stop or change direction within that distance. If it has to stop, it couldn't fly very fast at all, because breaking in mid-air is tricky. Instead, the Gabar Goshawk estimates the density and size of barriers in a landscape and then flies blind at the optimum speed that allows evasion.[80] Apparently, good off-piste skiers do the same thing, and autonomous drones are now being trained to imitate the birds.

LAPPET-FACED VULTURE
Torgos tracheliotos

VULTURE EYESIGHT AND HEARING

White the speed of visual perception is the superpower of pursuit predators, vultures like the Lappet-faced have long been assumed to have exceptional eyesight because they fly at extreme altitudes. Foraging vultures face major challenges in finding carrion: they live in big landscapes with relatively few carcasses, so they have evolved several tricks to help. First, vultures' eyesight is exceptional: their eyes have a higher density of light receptors than most birds and mammals, and they have an extra-sensitive patch that may increase magnification in the core area. This helps vultures see carcasses from around 2.7 times greater distance than we could.[81] But probably the most useful trick they have is watching their neighbours: each individual vulture patrols its own patch, but keeps a constant eye on its nearest few neighbours. If one bird starts moving with purpose, the others will follow. Using this network of neighbours, birds as far as 100 km away can quickly be attracted to a new carcass.[82] Recently we've learnt that vultures don't just look for carcasses and other scavengers, they listen too: when the calls of a distressed wildebeest calf, of feeding hyaenas and of lions were played at random locations in the savannah, nearly a quarter of the time vultures (often Lappet-faced Vultures) were the first scavengers to arrive.[83] This means that Lappet-faced Vultures not only lead mammalian scavengers to carcasses,[84] but sometimes they listen to mammals in order to find carcasses in the first place. It seems extraordinary that we've only learnt the importance of auditory cues for vultures so recently.

VERREAUX'S EAGLE OWL
Bubo lacteus

OWL HEARING

If the awareness of vultures' acute hearing is a relatively new insight, we've known for a long time that owls such as the Verreaux's Eagle Owl have exceptional auditory abilities. At 2–3 kg in weight, Verreaux's Eagle Owl is the largest of the African owls. It feeds mainly on nocturnal mammals. Like most owls, it is strictly nocturnal and can hunt in total darkness, pinpointing prey through hearing alone, though its eyesight in the dark is also great. To find prey only by hearing, owls compare both the difference in volume of sounds arriving at each ear and the difference in time it takes the sounds to reach them.[85] As most owls' left ears are higher than their right ears, they use volume differences to fix vertical location and time differences to find horizontal location, giving them the ability to pinpoint to 1° accuracy. Although mammals also employ both volume and timing to localise sounds, they only use one at a time, locating low-pitch sounds with time differences and higher-pitched ones with volume, and so they cannot achieve the same precision as birds. Owls also hear much quieter sounds that we can. Sounds are pressure waves in the air and we measure these in decibels (dB). The quietest most humans are able to hear is 0 dB; 10 db is ten times louder, while 20 dB (100 times louder) is my current limit with mild hearing loss. Owls can hear sounds as quiet as -12dB,[86] which is more than ten times quieter than the best human hearing. So, owl species such as Verreaux's Eagle Owls have amazing hearing, but their real superpower is an ability to turn that into 3D localisation of sound with accuracy. Using hearing alone, they would still fly into things in total darkness though, so huge eyes densely packed with light-sensitive cells are vital.

WHITE-BREASTED CORMORANT
Phalacrocorax lucidus

UNDERWATER VISION

Seeing in the dark is one challenge some birds have overcome; equally tricky is seeing underwater for species such as the White-breasted Cormorant. Like many waterbirds, it pursues fish in dives lasting around a minute. To do this successfully and also see well in the air is a major challenge. Our underwater sight is terrible because we rely on the curved, outer part of our eye, the cornea, to start the focusing process, with our lens doing fine tuning. The cornea does this because the speed of light in air is faster than in the cornea, resulting in the light bending. This process of refraction is what makes things seen through an air/water boundary appear closer than they are, as is easily observed by looking through a glass of water. However, because it mainly comprises water itself, our cornea bends light by about the same amount as water does and underwater pressure flattens our cornea, preventing us from focusing well. Cormorants have two ways to overcome this.[87] First, they use a clear, second eyelid called a nictitating membrane to provide structural support to the cornea, maintaining curvature despite water pressure. Second, and much more remarkably, they can push their lens out through their pupil and focus using the lens alone. In fact, the ability of birds like the White-breasted Cormorant to reshape their lens in order to focus is so good that they are able to focus on things five times closer than we can, in addition to viewing distant objects well. Together, these adaptations afford them good eyesight both above and below water.

AFRICAN DARTER
Anhinga rufa

UNDERWATER SWIMMING

A relative of the cormorants, the African Darter has taken underwater swimming to an extreme. Darters are unusual for waterbirds because they often swim half underwater, with just their long neck exposed,[88] giving them their alternative name of snakebird. Although cormorants can swim semi-submerged too, they need to keep moving so as to maintain neutral buoyancy, where darters can do this while stationary or moving very slowly. They can control their buoyancy because their breast feathers completely lack barbs to connect the feather edges, allowing them to absorb water and for the water to seep into all the air pockets below their feathers. Unusually for birds, darters also have largely solid bones and a reduced network of internal air-sacs – all features that increase their density. Most importantly, they actively control their buoyancy during foraging, expelling air from their air-sacs before diving and possibly even controlling the volume of air in sacs below their wings actively during dives. These features all allow darters to achieve neutral buoyancy, enabling them to slowly stalk potential food underwater, then stab and impale prey on their pointed beak in a way their cousins cannot achieve.[89] After a swim, however, African Darters have to dry their feathers, spreading their wings in a distinctive position with their backs to the sun that assists feather drying.[90] Surprisingly, although cormorants will do this when wet too, darters actually hold this position more often when they are dry, substantially reducing the metabolic costs of keeping warm.

WHITE-BACKED DUCK
Thalassornis leuconotus

DUCK WATERPROOFING

When it comes to waterproof feathers, the White-backed Duck is probably the opposite of the darter. White-backed Ducks are a widespread but rather scarce species of clean ponds and lakes. They feed mostly on submerged waterlilies and other vegetation at night, diving several metres down and capable of moving considerable distances underwater, but always popping back up completely dry. All birds regularly cover their feathers in preen oils, and oils repel water, so many people assume that duck waterproofing comes from the oils they use. However, duck preen oil is no different to that of other birds so can't explain the improved waterproofing.[91] Although preen oil does help generate waterproofing and ducks do produce more than many birds, the key to the superb water repellence of ducks is the interaction between the oil and the microscopic make-up of their feathers.[92] The structure of duck feathers lets them lock tiny air bubbles alongside each feather, preventing the water from making contact with most of the feather itself. These tiny bubbles are trapped among the barbules (the smallest substructure of the feather), which have tiny hooks to fix to the neighbouring barbules, creating a microscopic lattice of feather and hole. This structure generates considerable rigidity, which is important to stop water pulling the parts of the feather together; the oils keep the lattice repellent and the holes trap the air that prevents water getting close. So good at repelling water is the feather structure of ducks such as the White-backed Duck that people are now copying the design to develop new textiles that are a thousand times more water-repellent than traditional clothing.[93] Not only does this demonstrate how efficient evolution can be at solving problems, it also shows how curiosity-driven research can lead to valuable products no-one was expecting.

CAPE TEAL
Anas capensis

DUCK GENITALIA

Duck waterproofing is something we might want to copy, but the same cannot be said for the reproductive practices of species such as the Cape Teal. This is a widely distributed duck, found across much of Africa, particularly on alkaline soda lakes. Like most ducks, when they nest in higher densities things can get uncomfortable for females. 'Forced copulation' is the polite biological term for the mating behaviour of male ducks, and certainly happens in the Cape Teal.[94] Ducks living near each other tend to have a fairly strict pecking order, within which dominant pairs do pretty nicely, while the subordinates get a rougher deal. Although dominant pairs live relatively quiet lives, during the breeding season females in subordinate pairs are subject to persecution by dominant males. Male ducks try to force matings with females, sometimes in groups that are so violent the female can be drowned. To enable such forced matings, and unlike most birds, male ducks have a phallus that twists like a corkscrew. In species where males have a longer and more twisted phallus, the female's vagina is also longer and more twisted.[95] This might seem a necessary anatomical feature for females when males have a twisted phallus, but duck vaginas twist the opposite way to the male organ. In some species, females even have vaginas with dead ends. This ensures that if a non-preferred male successfully forces a mating, the female still has some control of paternity by forcing his phallus into a dead end. So a cozy-looking family of Cape Teal are possibly the result of forced copulations for the females, both with her social partner and with neighbouring males.

WHITE-FACED WHISTLING DUCK
Dendrocygna viduata

DUCK IMMUNITY AND DISEASE TRANSMISSION

Whistling ducks such as the White-faced Whistling Duck differ from most duck species by having very strong pair bonds and peaceful social lives, but they probably do share their general duck immune system that makes them significant in the spread of certain diseases. As with many wetland birds, White-faced Whistling Ducks are extremely widely distributed across Africa as well as in South and Central America. Living on temporary wetlands means long-distance movement is an essential part of their lives, but each movement risks transferring disease. Normally, we don't worry much about diseases in animals: of around 320,000 estimated mammal viruses,[96] only a tiny handful affect people, and the same is true for birds. But 70% of new human viruses do originate in animals,[97] including in ducks. In fact, ducks (and the better-known bats) are top of the list. Ducks have an unusual immune system that gives them exceptionally good antibody responses to common diseases and means they are able to behave normally, appear healthy and fly long distances while simultaneously shedding high viral loads, particularly for coronaviruses and influenza viruses. Add to the unique immune system the fact that ducks like White-faced Whistling Duck are highly nomadic, travel far while infected,[98] and often mix with domestic ducks at suburban wetlands and we see why monitoring disease in wildlife is so important. That species such as this have the potential to harbour diseases which could cross to humans shouldn't mean we target the birds, but does justify careful disease monitoring and suggests that better separation of domestic from wild animals is a really important aspect of disease control.

BLACK-WINGED STILT
Himantopus himantopus

FINDING EPHEMERAL WETLANDS

While ducks are capable of carrying diseases on their nomadic wanderings between temporary wetlands, it is species such as the Black-winged Stilt that are perhaps the experts in locating ephemeral wetlands. These elegant wetland birds are contenders for the award for the longest leg-to-body ratio of any bird and are common at temporary wetlands across Africa, Asia and southern Europe. When floods occur, they appear from nowhere within days, which raises the question of how they know the wetland has filled? Like many temporary wetland specialists, they have a challenge: some wetlands are very far from other suitable sites and only fill with water following exceptional rainfall. If they were to simply fly over to a known wetland site to check if it is wet, they could be in real trouble if there is no water when they arrive. The same is true if they were simply to wander around the landscape randomly searching for water. Alternatively, they could try to anticipate where floods will happen based on previous experience. From the rapidity with which these birds find newly flooded waterholes, we now know that African wetland specialists follow rain fronts as they cross the continent,[99] travelling along as the rains take them. But there's also evidence that at least some duck species use distant cues. Australian Grey Teal, for instance, have been observed to move hundreds of kilometres in a few hours, sometimes just to check if wetlands were filled (and some died when lakes were dry); and at other times they seem to use the smell of wetlands hundreds of kilometres distant.[100] So Black-winged Stilts probably stay at a wetland until it dries up, then they may pick up the scent of water hundreds of kilometres away, do a speculative dash to places they know are sometimes wet, and then actively follow cloud fronts: all the strategies seem to work.

LESSER FLAMINGO
Phoeniconaias minor

LIFE IN EXTREME ENVIRONMENTS

While the Lesser Flamingo tends to visit permanent lakes, its preference for very particular water depths and salinity make the population just as nomadic as ephemeral wetland specialists. Tanzania's Lake Natron is a nursery to nearly all of East Africa's population of 1.5–2.5 million of this species, and the sight of hundreds of thousands of them is arguably one of nature's great highlights. In the soda lakes of the Rift Valley, they live extreme lives. Lake Natron is a harsh environment: at over 30°C the 'water' is hot, and saturated with soda. It has a pH of around 10,[101] more or less the same as the ammonia you might clean a sink with, and it expands and contracts massively with the rains. So the flamingos that breed there need to be tough. Their beaks are evolved for filter-feeding on plankton-like *Arthrospira*, a cyanobacterium (single-celled organisms also known as blue-green algae) that thrives in extreme environments,[102] and that can generate anatoxin-a, a potent neurotoxin.[103] Such challenging environmental conditions mean that few organisms are able to live in Lake Natron, but for those that can, there is little competition. Water conditions change fast in these shallow lakes, and a flock of a million flamingos eat so much food that Lesser Flamingos need to keep moving. One individual fitted with a GPS tag made 44 different lake visits – from Bogoria in Kenya to Manyara in Tanzania – over 101 days, travelling a total of 4,700 km.[104] When conditions at Natron are perfect (only about once every seven years), nearly all the Lesser Flamingos of East Africa congregate here to nest: it is the only lake where conditions are ever sufficient to support so many birds for so long. Thus, to survive in such extreme environments the Lesser Flamingo has evolved a long life that ensures against regular failed breeding events, long movements to find suitable conditions and tolerance of toxic food.

BLACK-CHESTED SNAKE-EAGLE
Circaetus pectoralis

VENOM IMMUNITY

Flamingos aren't the only birds that eat food that kills other animals, and snake-eagles like the Black-chested Snake-eagle are probably the best example of this. Along with their close relatives, these birds are specialist reptile hunters, with a particular penchant for snakes – even the venomous ones. Eating a venomous snake is high risk: much as I love watching snakes, I keep my distance from venomous species. Snake-eagles, on the other hand, show discrimination only by size: the bigger the snake the better.[105] So, how do they avoid getting bitten? First, they kill fast and go for surprise. Second, they have very thick feathers on their bellies and thick scales on the legs,[106] giving mechanical protection equivalent to a person wearing thick boots in snake country. But most interesting is the suggestion that they may have some immunity to snake venom. There is only one study I know of that indicates this,[107] based on some rather unsavoury experiments with mice. Those mice injected with snake venom died, but mice injected with venom mixed with blood plasma from snake-eagles did not. Whether or not Black-chested Snake-eagles are indeed immune to snake venom, they're certainly very cool birds.

LITTLE BEE-EATER
Merops pusillus

VENOM REMOVAL

Like snake-eagles, bee-eaters such as the Little Bee-eater also snack on potentially dangerous prey. Little Bee-eaters are often seen feeding around mammals as they disturb insects, but unlike many bee-eaters they are not colonial, remaining as a pair or single family. As with most bee-eaters, they eat a wide variety of flying insects, but they will always bring bees and wasps back to a perch to remove the venom.[108] Experienced birds extract the sting by rubbing the tail of the insect on their perch until the sting comes off; young birds have to learn how to do this and can get stung while they are experimenting. Although they do watch their parents to learn to some extent, basic knowledge of both the species that have venomous stings which need removing and the process involved seems to be hard-wired. If they get things wrong, however, naive young bee-eaters can get stung. While this does seem to hurt them, they appear to have evolved a certain amount of immunity to the venom: a single sting would be fatal to most 30g birds, but bee-eaters can cope easily.

COMMON BULBUL
Pycnonotus barbatus

TOXIC BUTTERFLIES FOR DINNER

While bee-eaters are renowned for consuming toxic prey, the fact that Common (or Yellow-vented) Bulbuls do this too is much less known. As its name suggests, the Common Bulbul is one of the commonest and most widely distributed of birds in Africa, recently even establishing a toehold in southern Spain. One of the reasons it might do so well is its ability to eat toxic insects. The Monarch butterfly is known both for its impressive North American migrations, and for its caterpillars accumulating toxins from plants of the milkweed family. These toxins then serve to protect the insects from predation when they become adults.[109] Milkweed poisons make animal hearts beat stronger and stronger until this causes cardiac arrest, and most bird species vomit if they eat the butterflies. Not so the Common Bulbul, which is quite happy consuming them.[110] Some other birds such as quails eat Monarchs too, so it might seem surprising that the insects bother protecting themselves this way, if it doesn't stop them being eaten. This is an example of an 'evolutionary arms race', where the evolution of resistance in a predator is keeping up with evolution of toxicity in the prey: it doesn't look like a sensible defence now, but for the first toxic butterfly it was a great innovation, until the bulbul developed resistance. Of course, the toxins in Monarchs also protect them from other birds, which can only help, but the ability of Common Bulbuls to eat these butterflies without any bad effects and thereby monopolise a plentiful resource might be one of the reasons why they are such successful birds.

COMMON QUAIL
Coturnix coturnix

TOXICITY ACCUMULATION

Of all the birds that regularly consume toxic prey, the story of the Common Quail is one of the most curious. In Tanzania there is both a resident quail population and a migrant population that breeds in Europe. Wild quail eat a very healthy natural diet of seeds and invertebrates, and in turn, many people like to eat these gamebirds. But be careful: poisoning from eating quail ('coturnism' – from the scientific name *Coturnix coturnix*) is surprisingly common. As they migrate over the Mediterranean, Common Quail can eat seeds from plants that make them highly toxic (probably a species of woundwort,[111] plants in the mint family) and there is no way to differentiate toxic from safe quail: every bird eaten presents a game of chance. This has been known for a long time: the Bible tells us of Israelites dying of quail poisoning in Sinai, which was probably a result of coturnism. Migrating quail eat woundwort seeds, extract the plant's defence toxin, safely absorb these into their fat and then when people eat the birds, they get sick and die. We don't yet know exactly what the chemicals involved are, though we are aware that they must be fat soluble because chips cooked in quail fat have caused coturnism.[112] Only some people appear to be susceptible, but we have no idea why there is this difference between individuals. So, just like Monarch butterfly caterpillars, the Common Quail co-opts a naturally occurring toxic chemical in a plant to make itself poisonous to predators.

WILLOW WARBLER
Phylloscopus trochilus

EXTREME MIGRANTS

Alongside the Common Quail, a whole host of other birds migrate from Africa to breed in Europe, and my favourite is probably the tiny Willow Warbler. During the non-breeding season in Tanzania they weigh around 9 g, but as they fatten for their migration they easily achieve 15 g, becoming little butterballs. Then they fly up to 13,000 km to their breeding areas. Amazingly, we now have technology small and light enough to track them on their travels.[113] With tiny devices attached, three birds from eastern Russia were followed to their wintering areas in southern Tanzania. Their migration commenced in late August, and they went first north-west and west into Central Asia for five weeks, stopped there for about ten days and then headed south into sub-Saharan Africa. In Africa, they showed a series of stops for two to three weeks at a time, starting in Ethiopia, before moving further south and halting again for another few weeks. These movements probably correspond to birds shadowing the rains south, as Willow Warblers usually arrive in Tanzania with the short rainy season (and all the bugs that follow). Rather remarkably for such tiny birds, as well as migrating so far they somehow find the energy to completely replace their feathers twice a year – more than most birds do and an extremely energetically costly activity. So, Willow Warblers demonstrate how even the smallest of species can do amazing migrations, averaging up to 240 km per day for weeks at a time. Then in Africa they show what we now know to be one of several non-breeding strategies, that of 'serial residency', following seasonal resource pulses around the continent.

ABDIM'S STORK
Ciconia abdimii

INTRA-AFRICAN MIGRANTS

While Willow Warblers cross continents between breeding and non-breeding seasons, other common visitors to Tanzania such as Abdim's Stork are intra-African migrants, breeding in the Sahel and following the rains south. There are numerous species that migrate within Africa, but with only a few exceptions, we have little information about these migrations. As Abdim's Storks are large enough to wear satellite transmitters, we know more about their movements than we do most intra-African migrants.[114] In one of the only studies of its kind in Africa to date, several adult birds were tagged on nests in Niger. After breeding, they migrated east and a little south to non-breeding ranges near Lake Victoria, with one carrying on from there south to Zimbabwe. Individuals from as far east as Yemen do the same, mingling in Tanzania and southern Africa in the non-breeding season. The amount of mixing between nesting populations in the non-breeding season (termed migratory connectivity[115]), varies between species and may be important for conservation. With a high degree of mixing, habitat change in one non-breeding area impacts many populations, while populations of species in which distinct breeding populations are also found in different locations during the non-breeding season are more likely to fluctuate separately. In Abdim's Storks, birds nesting in the same village ended up in very different non-breeding sites, alongside birds from all over the breeding range. This is more mixing than we are aware of in many European–African migrants and hints at how different intra-African migrants may be; there is much more to learn here.

AFRICAN PARADISE-FLYCATCHER
Terpsiphone viridis

COMPLEXITY OF INTRA-AFRICAN MIGRANTS

As well as the general paucity of research in much of Africa, one of the reasons we know relatively little about intra-African migration of species such as the African Paradise Flycatcher is because their behaviour and movements are so complex. In Tanzania, we see African Paradise Flycatchers all year round and might be forgiven for thinking them resident, but this hides the subtleties of movement: seven subspecies live here at least some of the time, most thought to be migratory but all doing separate things. In the Udzungwas of southern Tanzania, African Paradise Flycatchers are absent from high forests during the cold season from February to August.[116] These movements are probably altitudinal migrations, relatively local to the mountain systems. Longer movements are represented by non-breeding visitors from South Africa, which most likely account for many of the widespread sightings during the dry season, while western Tanzania probably receives wet-season breeding visitors from farther west still. Nocturnal migrants recorded in Tsavo in southern Kenya in November may be non-breeding visitors from the Sahel,[117] while other populations may be genuinely resident. This complexity means that only very close study of marked individuals is likely to uncover the full details of migrations, and there is a good deal to suggest intra-African migrations may be rather different to the better-studied European–African systems. For instance, all migrants prepare for migration by increasing their body mass, but intra-African migrants such as African Paradise Flycatchers store energy for migration as extra muscle, while Palaearctic migrants store fat before migration.[118] Certainly, we still have much to discover on this topic.

EURASIAN BEE-EATER
Merops apiaster

EVOLUTION OF MIGRATION

A few species of migrants such as the Eurasian Bee-eater are both intra-African migrants and European–African migrants. Along with just a tiny handful of other species, Eurasian Bee-eaters have fully migratory populations that breed in Europe and North Africa as well as a separate migratory population breeding in southern Africa. The European population nests between May and August, while the (rather small) southern African population breeds from October to March. The few Eurasian Bee-eaters found in Tanzania in July and August are likely to be South African breeding birds.[119] Although little studied, the southern African population is genetically diverse and not separable from European populations,[120] which implies that there is regular immigration to this population from the European breeders. Despite this, the South African birds differ not only in migration pattern but also in the timing of moult. This suggests that migration is a fast-evolving character. Since the European–African migration system has had to evolve in the 12,000 years since the end of the last ice age, it seems probable that migrations evolve quickly, which is potentially good news for those concerned about climate change impacts on migration routes.

RED-BACKED SHRIKE

Lanuis collurio

MIGRATION TIMING

Although migration seems to evolve rather swiftly, research on the Red-backed Shrike has helped us understand one of the fundamental problems that migrants have to solve: knowing where they are. Red-backed Shrikes are long-distance migrants, breeding in Europe and spending the non-breeding season in south-east Africa. An essential part of finding their way on this route is knowing where they are at any given point in time. Knowing how far north or south you are is pretty easy: wait until the sun is at its highest point in the sky and the length of your shadow tells you exactly how far from the equator you are. But working out how far east or west you are is much harder, a challenge known as the problem of longitude. Human navigators solved this by building an extremely precise clock: if you know your latitude, what time it is at a known position (for example, in London) and what time sunrise happens wherever you are, then you can identify longitude. But your clock has to be accurate. Red-backed Shrikes do the same using an internal clock that is precise to three minutes per day.[121] How this clock works is still a mystery, but it involves the retina in the back of the eye, the pineal gland in the brain and a number of other anatomic features working in synchrony.[122] If we keep birds in unexpected cycles of light and dark, they get lost – suggesting that daylight is important for helping set the clock.[123] Yet although we know the Red-backed Shrike can tell the time precisely, we don't fully understand how.

AFRICAN STONECHAT
Saxicola torquatus

ANNUAL CLOCKS AND PHOTOPERIODISM

While migrants like shrikes have internal clocks accurate to within three minutes, some resident species such as the African Stonechat must be even more precise. This species uses day-length to determine when to nest and when to moult. That's easy where day-length varies a lot between seasons, but not in the tropics. At the equator, day-length of shortest to longest day varies by just two minutes. As birds typically compute season by relative differences in day-length one day to the next, it was long considered impossible for tropical species to use day-length to establish season. Thanks to work on African Stonechats we now know that this is incorrect: just like their Eurasian cousins, day-length is used to set annual cycles of readiness to breed and moult.[124] However, they won't actually start until the rainy season arrives.[125] Keep African Stonechats in an aviary with a constant twelve hours of day and night and they maintain accurate annual cycles for several years,[126] indicating that these birds have a strong underlying annual clock that needs only the tiniest of day-length clues to keep it synchronised. So, the African Stonechat demonstrates that not only do birds have daily clocks that are precise to a few minutes, they also have an annual clock synchronised by just seconds of day-length difference.

BLACK-SHOULDERED KITE
Elanus caeruleus

NOMADISM IN SAVANNAHS

Neither resident nor migrant, the nomadism shown by birds like the Black-shouldered Kite is an alternative lifestyle that works rather well in savannahs. Most birds of prey are territorial and have a traditional nest to which they return each year. Black-shouldered Kites are nomads, breeding sometimes twice a year,[127] and nesting in large numbers when small grassland rodents experience population booms. Savannahs are tough places: they can have extreme wet and dry seasons and very large variation in rainfall between years. This means food that is plentiful now may be scarce tomorrow – but a few wet years in a row could create a rodent bonanza, which is great for Black-shouldered Kites. During a rodent outbreak, this species can be ten times commoner in Serengeti than during normal rodent years.[128] The birds achieve this not by breeding fast but because they move to use the spatially and temporally variable resources common in savannahs, just as traditional pastoralist communities do with their cattle. So, Black-shouldered Kites show us how nomadism is a sensible strategy in variable environments. In northern climates, rodent outbreaks are often exploited by owls. While filling this niche in Africa, Black-shouldered Kites have evolved some very owl-like traits, including soft feather edges (to quieten their flight), nomadic behaviour and even a rather owl-like face pattern.[129]

SPOTTED EAGLE OWL
Bubo africanus

BROOD REDUCTION

One of the adaptations that species such as this have made to take advantage of highly variable food supplies is a breeding strategy known as brood reduction. It takes Spotted Eagle Owls quite a long time to raise young: around 32 days incubation and about 49 days to fledge. They eat small mammals that breed fast and vary greatly in population year on year, so the owls have to predict how much food will be available when the chicks are at their hungriest, several weeks after starting to breed. If they plan for a poor rodent year and lay only one egg but then it then turns out that there are lots of rodents, they would not do as well as a pair that had laid more eggs initially. But if they plan for a good year and then it turns out to be bad, the female will have invested energy in laying eggs and feeding young chicks that will not survive. Which is the better strategy depends on the relative costs of missing out on good conditions or wasting energy on speculative breeding. Female Spotted Eagle Owls always plan on having a big family, and lay one egg every couple of nights until they have a full clutch of three (some species, such as Barn Owls, follow the same strategy to even greater lengths and lay many more). Unlike most birds, owls start to incubate as soon as the first egg is laid, hence the chicks hatch one or two days apart, matching the laying interval. As young chicks grow fast, by the time the whole clutch has hatched there will be size differences between siblings. In a good year with lots of food all the chicks survive. In a bad year, the small chicks that hatched last are left to die. This – to us – rather callous strategy, called adaptive brood reduction,[130] ensures that owls are primed to exploit boom rodent years, but in most years the smaller chicks starve.

TAWNY EAGLE

Aquila rapax

SIBLICIDE

While owls will raise all their chicks when food is available, many species of eagle such as the Tawny Eagle lay two eggs, but always expect to raise only one. This is not due to lack of food but because these birds are 'obligate cainists':[131] one of the chicks will kill the other in the first few days after hatching. Because the first-hatched chick is larger, in practice it almost always kills the second. This seems wasteful: why should Tawny Eagles lay two eggs if only one chick survives? Moreover, if humans provide unlimited food to Tawny Eagles the older chick still always kills the younger. And if we separate the chicks so one doesn't kill the other, adults can raise both successfully – so the reason for this siblicide is not food limitation. Instead, we think that the second egg is retained simply as insurance against infertility and infirmity in the first chick. If the parents had laid only one egg and through some accident it was damaged or infertile, by the time this became apparent after a month of incubation there would be no time to start nesting again – wasting a breeding season. Furthermore, by letting young fight to the death, Tawny Eagle parents ensure the strongest is most likely to survive, which is important in the competitive world of raptors. Because there is nothing inherently wrong with the second chick in eagle nests, reintroductions of eagles and other species where siblicide is normal often make use of this second chick,[132] removing it at hatching and hand-rearing to provide extra young birds to reintroduce, while not impacting the source population.

CROWNED EAGLE
Stephanoaetus coronatus

EXTREME PREDATORS

While Tawny Eagles are impressive predators, my vote for Africa's most awe-inspiring carnivore goes to the Crowned Eagle. These birds mainly eat monkeys in the canopies of dense forests, which they catch in dramatic swooping ambushes. Nonetheless, when the opportunity presents itself they are quite happy to take mammals around ten times their own body mass, such as bushbuck, which they kill with the aid of a 10-cm-long, bacteria-drenched rear talon, puncturing vital organs and causing death by sepsis within a few hours.[133] This rear talon, the hallux, has two groves on each side that collect blood and tissue, providing a harbour for bacteria and forming a more impressive killing tool than the teeth or claws of any African mammal. Predation by Crowned Eagles on primates is extremely efficient (sightings of Crowned Eagles are the main source of monkey alarm calls[134]). Because they generally prefer smaller individuals, this predation exerts strong selection pressure among primates to grow bigger and forage lower in trees. There is even a theory that a Crowned Eagle relative forced our hominin ancestors away from an arboreal lifestyle, and may have encouraged the evolution of terrestrial behaviour. There is good evidence that our early ancestors *Australopithecus* were eaten by them,[135] and sadly (but very rarely), Crowned Eagles still kill children today: no other predator has a stronger claim to shaping human evolution.

MARTIAL EAGLE
Polemaetus bellicosus

MESOCARNIVORE RELEASE

The only other African eagle species in contention for best predator is the Martial Eagle. With a wingspan up to 2.4 metres, Martial Eagles are huge and can kill strong prey, including jackals and smaller cats such as servals. Top predators like these play several important roles in natural ecosystems. If we were to lose Martial Eagles, populations of the smaller predators that they eat (servals, caracal and lots of mongoose[136]) may increase. Medium-sized predators (mesocarnivores) consume lots of smaller animals and birds, so without top carnivores such as eagles, increased populations mesocarnivores can have serious impacts on the populations of the animals they eat. This 'mesocarnivore release' that can happen in the absence of top carnivores such as eagles can decrease rodent populations in turn altering the abundance of seeds that rodents eat.[137] Such changes can propagate through an ecosystem, such that even though top predators occur at relatively low numbers, they often have a disproportionate impact on the ecosystem,[138] and thus their conservation has to be a priority. So the Martial Eagle, as well as being one of the most impressive predators in the world, has a vital role to play in keeping savannahs the way we expect them to function. Happily, they're still relatively common birds in Tanzania, and always compelling to watch.

LONG-CRESTED EAGLE
Lophaetus occipitalis

PEST CONTROL

While larger eagles might have an impact on middle-sized carnivore populations, smaller species such as the Long-crested Eagle probably have a direct role in pest control that is beneficial for their human neighbours. Long-crested Eagles are a very distinctive small eagle, often found close to human habitations. The crest is eye-catching, and obviously the product of sexual selection, but it is the human association that interests us here. In Uganda, this is one of the commonest raptors in human-dominated landscapes,[139] so the fact that it mainly eats rodents suggests that this species has an important role in keeping down pests.[140] Weighing around 1 kg, it probably needs about 10 mice per day, so when nesting and feeding hungry chicks, a pair probably account for at least 30 per day (or three to four rats). That's a pretty useful pest control contribution in rural areas, so this is a bird that really should be looked after. As ever, the story is never totally simple: although Long-crested Eagles eat lots of rodents, it is really the number of rodents that determine the eagle population, not the other way around: what predators can do is reduce the peak abundance across the year, but they probably don't reduce the breeding population by much or they would have nothing to eat. Still, this snazzy bird illustrates the often-overlooked 'ecosystem services' that certain key species can offer.

NORTHERN FISCAL
Lanius humeralis

PEST CONTROL

Many birds, such the Northern Fiscal, eat mainly insects, and since a lot of insects are agricultural pests it seems reasonable to propose that, as with eagles preying on rodents, they may have a role in pest control. Fiscal Shrikes consume a lot of bugs: about seven large insect captures per hour in one study.[141] Smaller insect-eating birds generally have much higher foraging rates, but take smaller insects. So it might seem obvious that encouraging bird populations would be good for pest control. As ever, nature is never so simple: some insects eaten by birds might be doing important pollination jobs (looking at you bee-eaters), and many insects eat other pest insects themselves; the net costs and benefits of birds on farms in East Africa is therefore difficult to know. Happily, in an experiment on Kilimanjaro, coffee bushes were covered with cages that could selectively exclude birds, insects, or both, from coffee bushes.[142] This showed that removing birds alone reduced the number of coffee beans growing, while removing insects reduced the size of the coffee beans, and removing both was worst of all. Without insects, coffee had to self-pollinate, resulting in smaller beans; without birds, the insect pests damaged the plants. So overall, on a Kilimanjaro coffee farm, insect-eating birds such as Northern Fiscals increased coffee production by around 7% – quite significant for anyone who enjoys a cup.

EASTERN GREY PLANTAIN-EATER

Crinifer zonurus

SEED DISPERSERS

An important ecological role played by fruit-eating birds is as seed dispersers. Unlike their green, blue and red cousins the turacos, Eastern Grey Plantain-eaters are drab, and are common only in the most western parts of East Africa, where they live in riverine forest and mature gardens. Seeds of most tropical soft-fruit trees that fall immediately below the canopy have a less than 0.01% chance of germinating,[143] so they need to somehow move away from their parent in order to establish. Turacos eat a great deal of fruit, about 60 kg per bird per year,[144] and medium-sized birds like this can carry seeds a long way in their digestive tract before defecating. The coating of poo protects seeds from predators and is thus essential for maintaining riverine forest in savannahs like Serengeti.[145] Of course, ecology is never simple: by visiting lots of fruiting trees, Plantain-eaters are good at spreading plant disease too – including really significant ones like banana wilt, which is currently threatening global supplies (the bacteria that cause the disease can live on the beaks of these birds for up to five days[146]). So, Eastern Grey Plantain-eaters demonstrate how essential fruit-eating birds are for maintaining riverine forests (providing homes for loads of animals) but also show us that there are no free meals in ecology: with seed dispersal comes the potential for disease spread.

SCARLET-CHESTED SUNBIRD
Chalcomitra senegalensis

POLLINATION SERVICES

One of the most reliable ecosystem services provided by species like the Scarlet-chested Sunbird is pollination. Sunbirds are the African and Asian answer to hummingbirds (though they're not closely related: sunbirds are true songbirds, while hummingbirds are cousins to nightjars and swifts), feeding mainly on nectar and with brightly coloured males. But they're not just pretty: they're important too. We now know that 44% of plant species pollinated by sunbirds are used by humans as medicine, food, building materials or for some other purpose.[147] Although there are exceptions, it also seems that most plants are only visited by a few sunbird species, with relatively little overlap, so pollination by the right sunbird is vital. In nature, tubular red flowers like those of aloes are often bird pollinated.[148] Such red hues are conspicuous to birds but not so to bees,[149] making it the perfect colour to attract an avian pollinator and until recently ecologists often used the combination of flower colour and shape alone to guess the pollinator of a particular plant. We now know that such shortcuts are unhelpful, and only by detailed observation of individual species can we build up the full picture of pollinators. This is starting to show that while red, tubular flowers are often bird pollinated, sunbird pollination is actually much more important to a wider set of plants than previously thought, and particularly among some of those plant species important to humans.

GOLDEN-WINGED SUNBIRD
Drepanorhynchus reichenowi

BIRD TASTE

When species such as the Golden-winged Sunbird pollinate plants, they aren't doing it for fun but because the flowers reward them with nectar. These delightful creatures are one of my favourites, and are typical sunbirds of high mountains, feeding on nectar (and a few insects) among flowers of montane forest clearings. Nectar tastes sweet to us, but what about to birds? Although we've long known that birds reject distasteful food, because their taste-buds are not on their tongue like ours, but are hidden in the roof of the beak among saliva glands,[150] it is only relatively recently that we've realised birds have a reasonable sense of taste. For nectarivorous species like sunbirds, identifying sugars is important: some types of sugar are easier for them to digest than others, so discriminating between types of sugar is key and sunbirds have evolved dedicated sensors to do so.[151] As nectar from different plants differs not only in sugar types but also considerably in concentration and in the presence of trace amounts of salts, sunbirds also need sensitive taste for salts. Drinking dilute nectar presents a problem because if it is too dilute it will flush their bodies of salt: if local nectar sources are dilute they prefer saltier plants.[152] What no bird can taste is chilli: plants that only want birds to eat their fruits and disperse the seeds add capsaicin to the fruit to make mammals avoid them.[153] Hence, birds like Golden-winged Sunbirds have a better discrimination of sweet flavours than we do, but won't appreciate spicy food.

BRONZE SUNBIRD
Nectarinia kilimensis

STRUCTURAL COLOURS

The most striking thing about most sunbirds is, of course, the shining iridescence of their plumage. Like many sunbirds, male Bronze Sunbirds have beautiful shimmering feathers that change colour depending on the light. Unlike the reds and yellows of bird feathers that are made of pigments (rather like paint embedded within the feather), iridescence is in fact structural, formed by tiny holes in the proteins (keratin, like our fingernails) that make up the feathers themselves, and small structures called melanosomes that scatter light in an ordered and coherent way. Depending on whether the holes are small or large, spherical or tubular, and the precise shape of the melanosomes, different structures reflect different colours from the sunlight. Essentially, iridescence is created by the interference between waves of light reflected from two or more thin films: the edges of the hole and surfaces of the melanosomes, and this is the same process that generates rainbows on soap bubbles.[154] Most impressively, the melanosome and keratin compounds that generate structural colours within feathers can fossilise, and in this way we know that some of the feathered dinosaurs closely related to *Jurassic Park*'s Velociraptors also had iridescent patches.[155] By comparing structures within the feathers of the Bronze Sunbird and other shiny birds, we can thus reconstruct the colours of dinosaurs that died out millions of years ago! Elsewhere, evolution has led iridescent butterflies to converge on the same solution. Surprisingly, given how attractive they are, sunbirds are one of the few birds that look the same to humans as they do to other birds, as their plumage reflects remarkably little ultraviolet light (a key part of the palette for many birds).[156]

COLLARED SUNBIRD
Hedydipna collaris

SMALL BIRD METABOLISMS

As well as nectar feeding and iridescent plumage, sunbirds also share small size with the rather better-known hummingbirds. Collared Sunbirds are one of the tiniest of sunbirds and indeed one of the most diminutive of Africa's birds altogether, weighing around 7g. Being so tiny is a challenge: small birds require relatively more energy than bigger ones and are generally less well equipped to defend themselves against predators. Just staying warm is tricky for small birds, because keeping warm is harder with greater surface area relative to body core, so the relative costs of warming rise exponentially as body size decreases.[157] Usually, this means that tiny birds spend all their time feeding. Feeding more exposes you to more predation risk, which being tiny does too: even spiders catch sunbirds![158] For species where predation risk is high, there is little point saving energy for a future that may never happen – hence, small birds typically have many chicks, work really hard raising them and die young, while larger birds invest less energy in each nesting attempt, expecting to have more opportunities later. Sunbirds and hummingbirds are both exceptions here: Collared Sunbird pairs raise an average of 1.3 chicks per year, around one tenth of the expected productivity given their size.[159] They survive being tiny and having few young because the nectar they eat is a very high-energy fuel and they also drop their body temperature at night by 5–10°C. This means they meet their somewhat reduced energy demands in a shorter foraging time and can then hide from predators, prolonging their lives relative to similar-sized birds.

SPECKLED MOUSEBIRD
Colius striatus

THERMOREGULATION

Sunbirds aren't the only species that reduce their metabolic costs by lowering their body temperature at night, and Speckled Mousebirds are one of the best examples of this tactic. Extremely cute birds, mousebirds are now an exclusively African family, though they used to be much more widespread. Generally speaking, most birds maintain a strict body temperature of 41°C which, particularly for small species, requires lots of energy to maintain. Mousebirds are even more unusual because unlike most birds their size that eat high-energy insects or even nectar, they are herbivores, and feeding on plants doesn't provide fuel that is easy to extract. To save energy, they have also evolved the capacity to down-regulate their body temperature when roosting. At night Speckled Mousebirds reduce heat production and sleep in clusters, snuggling together to slow the decline in body temperatures overnight. Even given all the food they can eat, they will still drop body temperature to 33°C,[160] and in the morning they sun themselves in treetops to warm up. During the day they also climb to the treetops and sun their bellies, aiding digestion without increasing energetic costs. Lowering heat production but snuggling and sunning can save mousebirds 40–50% of energetic costs, moving a long way towards the energy-saving techniques of cold-blooded animals like reptiles. Other small-bodied herbivores such as hyrax use the same trick, sunning themselves on warm rocks.

AFRICAN SCOPS OWL
Otus senegalensis

TORPOR AND METABOLICS

One of the most unexpected species to alter metabolic rates so as to reduce energy costs is the African Scops Owl. The regular 'prrrp' calls of these common and widespread birds are great to hear in the African bush at night, but for owls they are truly tiny, weighing less than most thrushes. If you come across them during the day they are extremely reluctant to fly. This isn't simply because they trust their camouflage: although they are wonderfully cryptic, they clearly know when they've been seen. Instead it is because, like sunbirds, they often enter a shallow torpor when sleeping.[161] After a cold night, the owls reduce their body temperature by nearly 10°C to save energy the following day. Hence, African Scops Owls can reduce their energetic demands during the day if foraging is bad at night, though the cost they pay is being slow to fly away and thereby risking predation. Unlike mousebirds (where reduced heat production at night is independent of condition), African Scops Owls make a decision about whether or not to enter torpor based on their current state, weighing up the relative risks of starvation and predation. Of course, this is unlikely to be a conscious decision, but rather an instinct honed by generations of natural selection, but it does make them sleepy and very cute if you see them during the day. No doubt their cryptic plumage has evolved to reduce the predation risk, and given how many seem to be around and calling in some places at night, they clearly hide well during the day.

SLENDER-TAILED NIGHTJAR
Caprimulgus clarus

CRYPSIS

To my mind, the true experts when it comes to cryptic plumage are nightjars. While nesting and roosting on the ground among piles of dry leaves and twigs, the camouflage of the likes of the Slender-tailed Nightjar is exquisite. Individuals on nests will sit tight until approached to only one or two metres, so they need to hide really successfully. To make cryptic plumage work, of course, the birds need to choose nesting sites that match their plumage well, but to do this requires quite a lot of self-awareness. Slender-tailed Nightjars manage this though: from just the scattering of leaves and twigs within a few centimetres of the nest, to the colour of soil in the local landscape, birds select sites that most closely match their own patterns.[162] Moreover, those nests that better match their surroundings are about 30% more likely to fledge young than those that don't match so well – an extremely strong selection pressure rewarding better matching.[163] So, nightjars choose nest sites that match their plumage, and if they do this really well will be more likely to breed successfully, driving the evolution of camouflage. And those colours are absolutely beautiful up close too.

SPOTTED THICK-KNEE
Burhinus capensis

TASTY BIRDS

One of the main reasons why birds like the Spotted Thick-knee have evolved cryptic colouration is obviously because lots of other animals want to eat them. The Spotted Thick-knee is a nocturnal member of the wader family, though this species does not associate with water. They have the unenviable distinction of being the tastiest birds in Africa! We know this thanks to one of the stranger acts of British colonialism in the 1940s and 1950s, when one C. W. Benson, then of the Northern Rhodesia Game and Tsetse Control Department, shot and fed 200 species of bird to his staff after expert preparation by his wife.[164] The result of this extensive research revealed that cryptic-plumaged birds, like thick-knees, consistently taste better than conspicuous species and suggested that either being tasty means birds need to hide better from predators, or that non-palatable birds could benefit from signalling this through warning colours. This pattern, recently confirmed only to hold for females (presumably, bright coloured males have to risk predation to impress cryptically coloured females),[165] presupposes that bird predators share human taste. In a subsequent series of papers demonstrating the colonial government's commitment to thoroughness, the original authors then fed parts of the same birds to hornets, cats and rats, and ended up confirming that human taste preferences broadly reflect preferences of a remarkably diverse range of animal carnivores. This is an extraordinary story of colonial dedication to the unending pursuit of knowledge, but as usual with dubious research, history does not record the broader opinions of Benson's staff.

GREAT WHITE PELICAN
Pelecanus onocrotalus

GUANO AND COLONISATION

While colonialism has led us to discover some surprising things about birds, the Great White Pelican and its colonially nesting relatives had a surprising influence on the development of the colonies itself. These magnificent waterbirds live in dense colonies across Africa and parts of southern Eurasia, aggregating on islands in fish rich lakes. Where conditions are good, such colonies can be extremely large. As many as 10,000 pairs nested at Tanzania's Lake Natron in 1962, for instance, a population that would have eaten 3,000–5,000 tons of fish per year.[166] Necessarily, all that eating results in lots of excrement. Over the years, pelican (and cormorant) droppings at colonies can accumulate in impressive concentrations. Back in the nineteenth century, the nutrients in these bird droppings were very valuable. Local communities in Peru had used guano as fertiliser for a long time when the Prussian colonialist Alexander von Humboldt learnt about it at the beginning of the nineteenth century.[167] This led to a 'white gold' rush, fuelled by slave labour, and rapidly emptied the Peruvian guano islands. With unmet demand for more fertiliser, the search then turned to Africa and the Great White Pelican.[168] So, the quest for valuable poo opened the south-west coast of Africa to colonialism. Still today, the phosphate mine at Minjingo on the edge of Lake Manyara in Tanzania mines the fossilised deposits of a pelican and cormorant colony, producing some of the highest quality fertiliser in Africa.

NARINA TROGON

Apaloderma narina

COLONIALISM IN BIRD NAMES

Colonial legacies in African ornithology are often seen in bird names, including in the Narina Trogon, one of Africa's most beautiful birds. In possibly the most cringeworthy footnotes to colonialism, Narina Trogon is named for someone considered by François Levaillant to be the most beautiful African. Levaillant, who has a cuckoo, cisticola and woodpecker named after him travelled in southern Africa in the late eighteenth century. He collected birds for European museums and named several himself. Unusually for collectors of the time, he named two species, including Narina Trogon, after Black Africans. If Narina seems an odd name in this context, it is: Levaillant had a mistress from the local community whose name he found hard to pronounce, so he renamed her Narina, then (perhaps showing a little more sincerity) named the most beautiful bird he encountered in her honour. He also named the Klaas's Cuckoo for his African assistant. In doing so, Levaillant became the only colonial traveller to do this;[169] out of around 2,300 African bird species, many bearing human names, Narina Trogon is one of only two species named for Black Africans. Does it matter what birds are called? Well, I think so: naming promotes connection and ownership,[170] and conservation is more effective when communities feel a sense of ownership over the species they are protecting.

SPEKE'S WEAVER
Ploceus spekei

RACIST BIRD NAMES

While many of the human names that African birds have been given by colonial ornithologists are relatively benign, there are a few, such as Speke's Weaver, that commemorate more distinctly racist individuals. To my mind, the English explorer John Hanning Speke, for whom this stunning bird is named, captures the essence of everyday colonialism in East Africa. Known for 'discovering' Lake Victoria and searching for the source of the Nile, Speke spent most of his time describing the faces of the 'races' he came across. Among these, he described what he considered a single group of people that he deemed superior Africans and decided they were probably descended from the Biblical tribe of Ham, the lost tribe of Israel. In reality, the people he identified as descendants of Ham were a disparate group of tribes including the Tutsis of Rwanda that shared one thing: they generally had paler skin colour than their neighbours. There is, of course, no evidence for the 'Hamitic hypothesis': it was pure racism. But these divisions between ethnic groups and favouritism towards certain tribes sowed seeds that can be traced through colonial policies that ultimately caused the genocide in Rwanda in 1994.[171] So, while Speke's Weavers themselves don't care, they bear the name of a racist colonialist. By all accounts Speke wasn't the worst of the colonial ornithologists, but surely it is high time we changed a few African bird names.

YELLOW-BILLED KITE

Milvus aegyptius

WHY SPLIT SPECIES?

Changing the names of birds is not uncommon and while it can be done for matters of taste, in the case of species like the Yellow-billed Kite, the change reflects not preference but increased understanding. The Yellow-billed Kite has adapted well to living alongside humans and is common in many African cities. Not long ago it was considered a subspecies of the Black Kite that visits East Africa from Eurasia in the non-breeding season. So why do taxonomists split species? Sometimes splitting results from a debate about what exactly is a species. Some taxonomists prefer a species concept that says if two populations can produce fertile hybrids then they're not actually separate species; this is the biological species concept. Others define species by quantifying differences and applying a comparative test: if two populations differ by 3% of their genes, then they differ as much as humans from chimpanzees, and so must be different species. Yet splits often come from improved understanding of the relationships between recognised species, which is what has happened with the Black Kite and the Yellow-billed Kite. Two species of kite live in Europe: Black Kites and Red Kites, the second a distinctive species not found in Africa (though like many good species they do sometimes hybridise[172]). Yellow-billed Kites look most like Black Kites, but Red Kites are quite different, so based on plumage differences Yellow-billed and Black Kites were assumed to be close relatives, with the Red Kite a more distant cousin. But with modern genetics, we now know that Red Kites and Yellow-billed Kites are actually the closer relatives.[173] As there is little doubt that Black and Red Kites are separate species, nesting alongside each other in much of Europe, and since Yellow-billed are obviously different to Red Kites, in the end it seemed sensible for taxonomists to split all three. Such 'hidden' species may be much commoner than we think.

BEESLEY'S LARK
Chersomanes beesleyi

WHAT'S IN A SPECIES?

Usually the species concept and exactly where we draw the line between species and subspecies is largely academic, but for a few very rare populations like that of the Beesley's Lark, such distinctions may be of considerable conservation interest. Beesley's Lark may or may not be one of the rarest bird species in Africa, with no more than 200 individuals living on one nutrient-rich plain north of Mt Meru. There is no doubt that Beesley's Lark is related to the more widely distributed Spike-heeled Lark. Beesley's is generally smaller and shows more differences between sexes than Spike-heeled, but these plumage differences are slight and overlap with some of the variation known within Spike-heeled.[174] Looking at the genetic data available, Beesley's is estimated to have separated from Spike-heeled Lark around three to five million years ago,[175] longer than many pairs of lark species that are widely accepted as being different. Most taxonomic authorities recognise this split, but not all. Unfortunately for Beesley's Lark, one of the authorities that does not consider this bird as separate from Spike-heeled is BirdLife, the main international organisation that assesses conservation risk in birds. BirdLife suggests it will not make a decision on species status until genetic evidence is available from a Spike-heeled population in south-west Democratic Republic of Congo. If those data meet expectations, Beesley's Lark is an extreme rarity, urgently needing protection. As things stand, it seems likely that this dainty lark will vanish before anyone can sample DNA from DRC.

YELLOW-COLLARED LOVEBIRD
Agapornis personatus

SEXUALLY SELECTED SPECIATION

Where different populations of a bird occur over large geographic ranges, each slightly different from the next, species limits can be particularly hard to identify – as illustrated by the sumptuous Yellow-collared Lovebird. Lovebirds are small, seed-eating parrots found only in Africa. They're well known as pets, and several species hybridise quite freely, including in the wild. Such prevalence of hybrids implies that the species in this group are closely related: indeed, given how freely they interbreed, some would question whether they really are different species at all. The latest genetic evidence suggests that one of the first lovebirds appeared in Africa's forests, where the Black-collared dwells today.[176] Fairly soon, a population switched to eating seeds and moved into the grasslands. Since then, five closely related species have split as they colonised different parts of the semi-arid corridor that surrounds West and Central Africa's lowland forests: first Peach-faced, in Namibia, then Lilian's and Black-cheeked in Zambia and Malawi, before Yellow-collared and more recently, Fischer's, both in Tanzania. Although in the wild some neighbouring populations do interbreed, given free choice and plenty of potential mates, males and females of each population choose each other – supporting their identification as different species. Assortative mating within populations indicates that genes generating preference for particular plumage patterns and the genes that generate those patterns are closely associated; these are the same conditions that underpin runaway evolution of excessive tail-lengths in some African bird species. The rapid generation of different, but very closely related species in lovebirds is therefore probably driven by sexual selection, resulting in strong plumage differences between species but little genetic differentiation – allowing occasional gene-flow from one species to another.

GREY-BREASTED SPURFOWL
Pternistis rufopictus

HYBRID SPECIATION

Hybridisation between closely related species is common in birds: more than 15% of species are known to hybridise in the wild.[177] As well as confusing ornithologists, sometimes such events can generate entirely new species, and one of these may be the origin of the Grey-breasted Spurfowl. This bird is only found within the Serengeti ecosystem, where it comes into contact with Yellow-necked and Red-necked Spurfowls at the edges of its range. Quite different to both of those species, recent genetic analysis suggests it may be of hybrid origin. Several obviously different spurfowl species are known to hybridise occasionally,[178] though most individuals of a species mate with their own kind and hybrids are often sterile or of reduced fertility.[179] Creation of entirely new species through hybridisation is rare in birds (though much more common in plants),[180] but nonetheless we now know that such events may be important in evolution, allowing one species to borrow genes from another. Indeed, genomes of most humans outside Africa contain evidence of our own ancient interbreeding with Neanderthals,[181] from whom Eurasians have inherited various genes, including many involved in disease resistance.

BLACK CUCKOOSHRIKE
Campephaga flava

AUSTRALIAN EMIGRATION

One of the more remarkable pieces of information we have learnt from modern genetic studies is the surprising importance of Australia in the evolution of birds like the Black Cuckooshrike. This intra-African migrant is part of a group of birds called the Corvoidea which originally evolved in Australia. How they got to Africa is rather surprising. True songbirds, known as passerines, have their evolutionary origins in Australia, and a pattern of repeated waves of species dispersing out of Australia and through Asia to colonise the rest of the world is well established.[182] The cuckooshrikes and their relatives represent one of these emigrant waves. While most of the emigrant waves seem to have dispersed north into Asia and from there west and then south to Africa, rather unexpectedly, the best evidence suggests that cuckooshrikes have repeatedly dispersed across the entire Indian Ocean,[183] direct to Africa. The patterns of relationships within the cuckooshrike family implies that they have achieved this massive intercontinental dispersal not just once, but several times. The Black Cuckooshrike and its close African relatives comprise one such group, whose ancestors made the journey around 12 million years ago. The Grey Cuckooshrike ancestor crossed earlier, about 15 million years ago, and the first group around 25 million years ago. Since their relatives are capable of crossing the Indian Ocean, it is perhaps not a great surprise that Black Cuckooshrike movements are a bit of a mystery. But I do find it astonishing that birds like this can make direct crossings over 10,500 km from Australia to Africa, in sufficient numbers to found new populations at the other end. And why Australia should be so productive in the evolution of many songbird families is even more of a mystery.

GLOSSY IBIS
Plegadis falcinellus

INTERCONTINENTAL MIGRATIONS

While intercontinental colonisations of birds are relatively rare, even today we know of several species – such as the Glossy Ibis – that have successfully completed long sea crossings, and waterbirds are one of the best in this respect. The first breeding record for Glossy Ibis in North America occurred in Florida in the 1880s,[184] before the species expanded north and southwards. Since the year 2000, a total of four birds ringed in Spain have been seen over 7,000 km away in the Caribbean, showing continued westward movement. Cattle Egrets have made the same transatlantic trip even more recently. In fact, many African waterbirds have very large ranges compared to similar-sized species that depend on other habitats. One possible explanation for this relates to the temporary nature of many African wetlands, which forces mobility on the birds that use them. That African birds have colonised the Americas seems amazing, but looking at close relatives indicates that it has actually happened frequently throughout history. Almost all the intercontinental colonisations between Africa and the Americas seem to have occurred in that direction, with no definite cases of American colonisation in Africa. The most likely explanation for this is because the Trade Winds blow to the west, which probably also assisted the fewer intercontinental avian colonists from Australia to Africa.

COMMON OSTRICH
Struthio camelus

THE GIFT OF FLIGHTLESSNESS

Until very recently, the distributions of the two African species of ostrich (the other being the Somali Ostrich *Struthio molybdophanes*) and their flightless relatives the ratites were something of a puzzle. The best explanation for their current distribution was considered to be a flightless ancestor that was widespread on Gondwanaland, the earth's single ancient southern continent that split up around 180 to 100 million years ago to form the modern continents, taking these early flightless birds to each part.[185] Thanks to modern genetic methods,[186] we now think that – far from being relics of Gondwanaland – relatives of ostriches (in Africa), the Emu and cassowaries (in Australia), kiwis (in New Zealand) and rheas (in South America) flew to their continents and then each lost the power of flight separately. There are two lines of evidence to support this: first, using the rate of evolution to date the last common ancestors among this group we know that the separation of the species is much more recent than the separation of the southern continent. Second, another member of the ratite group, the tinamou from South and Central America, can still fly, but instead of being the oldest split from the group (allowing flightlessness to evolve once and then be inherited by the rest), they are the most recent split. This means that either early ratites evolved flightlessness once, and then the branch leading to tinamou re-evolved flight, or (much more likely) ancestors flew to each continent and then evolution favoured loss of flight several times in this group: very unusual! Being flightless frees ostriches to grow as big as they want (over 100 kg) and doesn't stop them being fast, with a top speed of around 60 kmh.[187] Of course ostriches are interesting in many other ways too: they have a penis,[188] males do most of the care for chicks and will steal other those of other males,[189] thus increasing their parental responsibilities, and they probably have the largest eyes of any land animal (bigger than their brain!).[190]

BLACK-FACED SANDGROUSE
Pterocles decoratus

PRECOCIOUS OR ALTRICIAL SPECIES

Ostrich males steal other male's chicks because this dilutes the chance of their own young being eaten and comes at no additional cost: the chicks are able to feed themselves from day one, a trait shared with many other bird species including the Black-faced Sandgrouse. These lovely creatures are common in bushed savannah and have extremely cute fluffy chicks. Leaving the nest within hours of hatching, these chicks are precocial, while other birds have helpless chicks. Why should there be this basic difference between precocial and altricial (helpless) chicks? In reality, there is a bit of a spectrum, with some chicks being intermediates between the two. Altricial young require parents to bring food throughout the nesting stage; precocial chicks can feed themselves very soon, and a few super-precocial chicks such as those of the Australian megapodes can even fly from hatching. Evidence suggests that the first birds (and many other dinosaurs) were precocial.[191] So, having altricial young has evolved independently several times in birds.[192] My life as a parent would be much easier if my children needed little care, so why should birds evolve a requirement for increased parental care? First, eggs of precocial young have more yolk of higher calorific value than those of altricial young, and are therefore more costly to produce.[193] Precocial birds often compensate for these increased costs by having smaller clutches (though clutch size increases with body size, the exact opposite of what we see in altricial species). Cheap, plentiful eggs are a sensible strategy when food supply is predictable, so birds with precocial chicks are overall rarer in the tropics than elsewhere, but these patterns are not easy to explain. In general, we think that having altricial young may be energy saving, explaining its repeated evolution.

TEMMINCK'S COURSER
Cursorius temminckii

FIRE IN THE SAVANNAH

The precocious young of the Temminck's Courser are coloured perfectly to hide on recently burned ground, reflecting this attractive wader's close association with grass fires. For many visitors to African savannahs from temperate zones, discovering that rangers in savannahs may burn grasses every year or two (in some areas even more) is shocking. But actually, recent declines in fire are worrying: savannahs need fire.[194] Species such as Temminck's Courser flock to burned areas, often following smoke to find them. They have specially adapted blackish eggs for camouflage on scorched backgrounds, and their chicks are darker than most.[195] Such evolutionary adaptations in birds, and the ability of savannah plants to resprout tells of long fire histories, but people often ask if savannah fires are natural, since humans start almost all of them. To me, this question reflects an assumption that savannahs should be empty of people, ignoring the fact that humans have co-evolved with the savannah landscape. Our ancestors were lighting fires in savannahs more than a million years ago, before most of the animals and plants that live there today had evolved.[196] Rather, removing human-set fires would be 'unnatural'. Species such as Temminck's Courser and other fire-adapted birds, plants and animals reveal the ways in which modern savannahs have coevolved with fire-lighting humans. While we might question exactly how and when these fires should burn to best achieve our goals, removing them would be a bad mistake.

YELLOW-THROATED SANDGROUSE
Pterocles gutturalis

ARID ADAPTATIONS

While fire-adaptation may seem extreme, the adaptations of Yellow-throated Sandgrouse for living in arid environments show that fires may not be the biggest challenges in savannahs. To take advantage of temporally and spatially variable resources, this species is one of the true nomads of the bird world, appearing anywhere in East Africa and superbly adapted to the challenges of dry environments. Sandgrouse can live in deserts where daytime air temperatures exceed 50°C.[197] Because they eat mainly dry seeds, they need to make daily round trips of up to 160 km to find water.[198] That they survive at all seems a near miracle! Precocial as their chicks may be, they obviously cannot make long trips to waterholes but still need to drink (and the eggs must be kept moist), so male Yellow-throated Sandgrouse have specially adapted belly feathers that let them carry water to their young – these feathers are about twice as absorbent as a sponge.[199] Their arid adaptations don't stop here: birds can't sweat, so when they get hot most gape or pant to get cool at moderate temperatures. By contrast, at 40°C sandgrouse fluff their feathers up to increase insulation, and only when the temperature hits 50°C do they start gaping! They also have kidneys that recycle salt.[200] If you get a chance to watch them at a waterhole you'll also notice that they're rather beautiful and that unlike most birds, except their near relatives the doves, sandgrouse suck: they don't need to lift their heads each sip to drink.

VON DER DECKEN'S HORNBILL
Tockus deckeni

BEAKS AS HEAT SINKS

No bird can sweat, but methods for keeping cool vary considerably between species. For me, one of the most surprising ornithological discoveries of the last few years has been that hornbills such as the Von der Decken's Hornbill use their big beaks as radiators to control body temperature. The advent of thermal cameras led to the discovery that as air temperatures approach bird body temperature, some species with large bills divert blood to their beaks to dissipate excess heat. A toucan sheds 60% of its heat through the beak,[201] and although they're not closely related, the same thermal exchange is also true for hornbills.[202] Toucans and hornbills are good examples of convergent evolution: both are fairly omnivorous, have huge beaks and use what is called 'ballistic' eating mechanisms, throwing food down their gullet, but hornbills are more closely related to hoopoes and wood-hoopoes, with toucans relatives of barbets. Ornithologists hadn't considered beaks as important for temperature regulation before, because they seem hard and inert. But this isn't the case: colourful sheaths like those of Von der Decken's Hornbill need ample blood supplies to grow, and beak sheaths keep growing like fingernails throughout a bird's life. So the likes of Von der Decken's Hornbill have repurposed for use as a heat sink beak architecture that probably initially evolved to enable fancy displays, converging on similar designs and similar secondary functions to the distantly related toucans of South America.

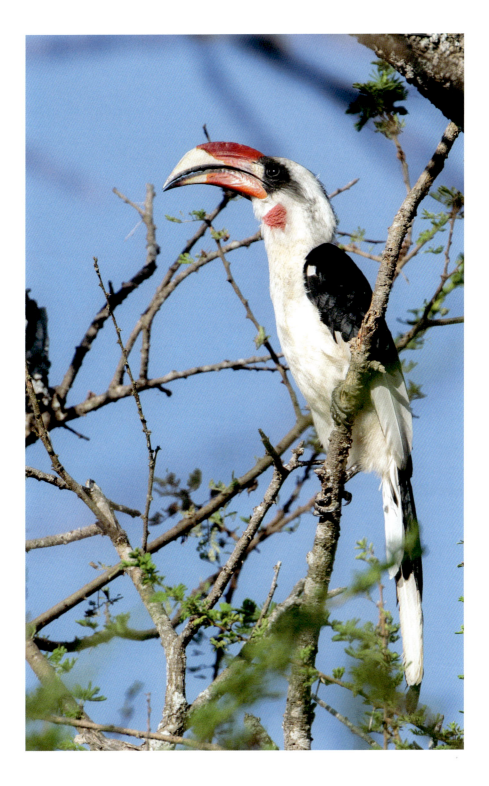

RUAHA RED-BILLED HORNBILL
Tockus ruahae

NEST BEHAVIOUR

Probably the most unusual feature of many hornbill species like the Ruaha Red-billed Hornbill are their nesting habits. This species is only known from Tanzania, occurring in central Tanzania as far north as Serengeti, and like most hornbills at nesting time females are sequestered in nest holes for long periods. After mating, females store sperm in their reproductive tract and use mud to seal themselves inside a tree hole, leaving only a crack wide enough to poke their beak through. The male will then feed her through the hole, and over the course of a few days she will fertilise and lay her eggs. When the clutch is complete, she then moults all her flight feathers at once and starts to regrow the feathers, while incubating her eggs. Best studied in the related Monteiro's Hornbill of Namibia,[203] this behaviour is extremely unusual for birds: males alone feed females (and later the chicks) for up to 70 days, during which time the females are completely dependent on their partner, as although they could break the nest seal, they cannot fly until their flight feathers have regrown. There are numerous theories about why such a strategy may have evolved, but the best evidence suggests it is driven by the female maximising her interests over that of her chicks (which would develop faster with her helping feed them), in exchange for promising perfect fidelity to her mate.[204] So, female Ruaha Red-billed Hornbills are faithful to their partners because they are isolated from the rest of the world, but they're prepared to sacrifice their children while doing so – proving once again that perhaps birds should not be a model for human behaviour.

HAMERKOP

Scopus umbretta

NEST MADNESS

If hornbills have unusual nesting behaviour, the Hamerkop has a complete nesting fetish! This bird is the sole representative of its family, most closely related to the Shoebill and pelicans, and like them a wetland species. Weighing about the same as a large pigeon, Hamerkop are compulsive builders of huge nests. Each pair constructs between three and five nests per year, and each nest is truly huge.[205] Most extend to more than 1 m in at least one dimension and require over 8,000 deliveries of branches, grass and mud to create,[206] with the female doing most of the work. To make one such nest a year would be impressive, to build multiple is truly absurd and raises the question, why? Such a nest fortress may seem to guarantee the safety of the contents, but Hamerkops actually have rather low nesting success, abandoning many clutches or failing with chicks later (often because eagle owls take over their very attractive real estate[207]). On average, each pair succeeds in fledging fewer than one chick per year. So the nest doesn't offer particularly good defence, and indeed it seems to attract unwanted usurpers. It does, however, provide a temperature-controlled environment (at a very constant 30°C) which may be important.[208] Yet most explanations suggest that building massive nests is not a functional requirement, but instead a pair-bonding ritual: it is good for couples to have a shared interest and with Hamerkops, compulsive home building seems to be the thing.[209]

CARDINAL WOODPECKER
Dendropicos fuscescens

HOLES AND ENVIRONMENTAL ENGINEERS

While Hamerkop nests often end up providing homes to owls and Egyptian Geese, woodpecker holes such as those drilled by Cardinal Woodpeckers offer dwellings to a huge range of other species. This species is the smallest common woodpecker in East Africa and they are widespread in many wooded sites. Like other woodpeckers, each breeding season a pair of Cardinal Woodpeckers will excavate a new nest cavity, taking a few weeks to complete the hole. Once the birds have finished nesting, which takes about two weeks, these holes then function as important refuges to a wide range of species – giving woodpeckers status as chief ecosystem engineers of the bird world.[210] Ecosystem engineers are animals that modify the landscape in important ways, and thereby create opportunities for a broad variety of other species. Elephants are a typical mammalian example, excavating waterholes and modifying tree cover. In East Africa, old woodpecker holes are often used by lizards, bats, galagos and tree rats, as well as hornbills, barbets and owls. So having woodpeckers in a landscape enables diverse ecosystems. All the pecking required to chisel out a suitable hole requires serious adaptations: a woodpecker excavating a hole experiences deceleration in its head that is equivalent to breaking from 42,000 kmh to stationary in 1 second (1,200 G).[211] Humans develop dangerous concussion at 80 G, so these are extreme birds – and they minimise brain injuries through a number of anatomical innovations,[212] most important of which is a specialised structure that wraps the tongue all the way around the brain like a seatbelt. This absorbs and displaces the worst of the impact shock.

AFRICAN SPOONBILL
Platalea alba

BILL FUNCTION

Woodpecker bills may be built for strength and durability, but the African Spoonbill's beak is equally specialised, if rather crazy looking! Spoonbills are relatives of storks and ibis, both of which have relatively normal beaks, so the spoon shape is an innovation that has evolved specifically in the spoonbills. So how does it work? When feeding, African Spoonbills hold their beak open a fraction and swing it from side to side, often in groups acting in synchrony. All around the edge, spoonbill beaks are embedded with tiny sensory pits like those of ibis and other birds that probe in mud, making them very sensitive to touch.[213] This probably enables them to rapidly sense animals moving in the water next to their bill, so they can catch them when sifting. Remarkably, there's also some evidence that the spoon shape and sideways movement that are so much a part of their feeding method generates a small vortex, like an upside-down whirlpool, which can suck small animals off the surface of muddy bottoms and into their beak, where the sensitivity means they will easily snap up the food. Although this idea is nice to think about, there's little proof that it works in nature, and since much of the spoonbill diet is made up of frogs (which are probably too heavy to move in this way) and aquatic invertebrates that swim freely,[214] it may simply be an example of anatomists getting over-excited about theoretical possibilities.

SACRED IBIS
Threskiornis aethiopicus

CULTURE AND HISTORY

A close relative of the spoonbills, the Sacred Ibis is remarkable not so much for its beak as for its role in cultural history. Sacred Ibis are common across African wetlands, and were once even more widespread. Although they have not actually bred in Egypt since around 1850, the species gets its name from the Egyptian god Thoth to whom it was sacred. For the ancient Egyptians, Thoth was the god of wisdom, knowledge and writing, and was also the herald of the Nile's annual flood.[215] Why there should be such an association is unclear: perhaps migrant ibis arrived with the annual Nile flood? What is certain is that to some ancient Egyptians, Sacred Ibis were an obsession of sorts: millions were sacrificed and mummified – at one site alone archaeologists have uncovered over four million ibis mummies![216] Some of these were taken to Europe by Napoleon in around 1810, where ornithologists in the French national museums studied the bones and concluded that they were the mummies of Yellow-billed Stork, which still carries the scientific name 'ibis'. The identification sparked debate, because although the bones were similar to those of storks, they were smaller and the beak much more curved, implying that in the 3,000 years since the birds were mummified, evolution had occurred. This occurred decades before Darwin, who identified the mechanism by which evolution occurs but not the original ideas itself. After a short period of debate, however, Georges Cuvier, an exceptional anatomist also working in France, studied the mummies and realised that they matched those of a species in the museum that had not yet been given a name. He thought this was evidence that evolution didn't happen,[217] because if Sacred Ibis had not changed in 3,000 years, there was no reason to believe species would alter over any length of time. Since Cuvier had so neatly identified the species and was in other ways a brilliant biologist, his argument against evolution was valued and there is a very strong case that the Sacred Ibis had an important role in delaying the development of evolutionary theory until Darwin arrived decades later.

BATELEUR
Terathopius ecaudatus

COLOUR MORPHS

We know that evolution selects between variations within a population but for some species, such as the Bateleur, obviously different variants exist at an apparently stable frequency. This colourful and uniquely adapted relative of the snake-eagles is widespread but fast declining across much of Africa. Adults have a panel of feathers on their back that comes in two different morphs, the more common chestnut variant and a rare cream colourway. In fact, Bateleurs are not the only African raptor to have colour morphs, and the reasons why these occur are probably varied. Some variations are themselves adaptive, each morph better suited to slightly different conditions. Black morph Great Sparrowhawks, for instance, seem to hunt more efficiently in shadier forests, while white morphs do better in bright sites.[218] Dark morphs of many raptors are commoner in damp forests, where the extra melanin in dark feathers may protect them from fungal infection.[219] But in the case of the Bateleur and many other polymorphic birds, direct benefits in different places appear unlikely. For these species it seems more probable that the gene altering colour is located on the bird's DNA very close to another gene that is being held in balance by evolution: for example, in owls, grey morphs are often more temperature sensitive than brown morphs by way of such genetic linkage.[220] So the colour morphs of birds like the Bateleur are unlikely to affect survival directly, but may be linked to another gene that is under selection in different ways in different environments. Cream morphs are slightly commoner in drier landscapes, but beyond that, the reasons two morphs persist is a mystery.

SECRETARY BIRD
Sagittarius serpentarius

HUNTING ADAPTATIONS

While Bateleur are unusual raptors, the prize for most off-beat African bird of prey surely belongs to the Secretary Bird, the only member of its family. Its unique approach to hunting allows it to tackle venomous snakes and might tell us about how extinct Terror Birds once hunted. Secretary Birds stalk the plains, looking for reptiles and insects, which they kill with a rapid kick, stamping them to death. Weighing 4kg, their kick generates a force several times this,[221] for just over a hundredth of a second. Such speedy kicks are much faster than the bird can consciously control, so they must be directed and then fired off without further adjustment, like shooting a gun. Given the potentially lethal prey they are taking (there's no evidence that Secretary Birds have immunity to snake venom), they need to be really accurate. But it is the force of the Secretary Bird's kick that is awesome, and it needs putting in context by reference to footballers: professional soccer players kick balls with a force equivalent to about twice their body mass. Secretary Birds kick at about five times that power, so while the likes of Lionel Messi may be able to kick a football 80 metres, a human-sized Secretary Bird might manage 400 metres! Seeing as the extinct Terror Birds of North America were pretty much scaled-up versions of Secretary Birds,[222] just imagine how a 300 kg version would kick.

AFRICAN OPENBILL
Anastomus lamelligerus

SPECIALISATION

Secretary Birds have developed a particular hunting technique that works well for them, but they're far from the only seriously specialised bird in Africa. The African Openbill is a stork that has a true obsession with eating water snails, which is facilitated by having evolved a beak that remains open by 5–6 cm in the middle when the tips are fully closed. So how do they use these beaks to eat snails? First, the gap appears as an adult; it isn't present in juveniles and therefore can't be essential for foraging, but nor is it created by a process of wear. Instead it develops through a process of delayed growth to ensure that when the beak is open, the tips are parallel to one another.[223] This helps the bird hold a hard, round object (a snail) between its bill tips, much like tweezers slope inwards at the end. Openbills have another specialist bill adaptation too: the inside of the tip is covered in brush-like ridges that both enable easier holding of shells,[224] and probably also help cut the snail's attachment to its shell – when an openbill finds a snail it extracts the animal from the shell by forcing the lower mandible under the hard covering that snails use to shut themselves in, then cuts the connecting muscle with its bill tip. Specialist feeding mechanisms enable birds like African Openbills to monopolise often plentiful resources without competition from other species. The cost, of course, is that if something were to suddenly reduce snail abundance, African Openbills would have little alternative to fall back on, suggesting that extreme specialisations should only be found in environments that have been stable over a long periods.

WHITE-CROWNED LAPWING
Vanellus albiceps

ARMAMENTS

Specialised feeding mechanisms are not the only reasons for birds developing unusual anatomical features, of course, as illustrated by the White-crowned Lapwing. This rather attractive species lives on large southern African rivers and both sexes have impressive wing spurs up to 15 mm long. Wing spurs have evolved independently in several groups: the spurs of the lapwings are bony extensions of their wrist covered in keratin horn, while those of the Spur-winged Geese, for example, are growths from 'hand' bones.[225] The White-crowned Lapwing probably sheds and regrows the keratin sheath each year. Armaments like these spurs are obviously used for combat, and a constant source of conflict in the bird world is males fighting for access to females. Such competition could cause sexual selection, but when birds have weapons, both sexes carry them – so, although fighting males do use spurs on each other, this is unlikely to be their primary role. The same is true for mammals: many antelopes have horns in both sexes and deploy them for defence, a few have them only in males and use them in male-on-male fights. In fact, having weapons primarily for mating fights is rare in nature,[226] because few animals want to risk harm. Therefore, the primary reason for wing spurs is probably nest defence. It might seem that the spurs of birds such as the White-crowned Lapwing are too small to be effective, but combined with an aggressive attitude that belies their size, inflicting a little damage may scare off much larger beasts than you expect.

AFRICAN WATTLED LAPWING
Vanellus senegallus

WHY WATTLES?

While some lapwings have spurs, the African Wattled Lapwing has gone in for wattles in a big way. This is a widely distributed bird of wet patches in savannah, and rather handsome too. Although the African Wattled Lapwing is named for this specific feature, many lapwing species have wattles and to me they look rather comic. But what is their purpose? In general, wattles probably serve three main roles: as a sexually selected ornament, as a signalling organ and to assist thermoregulation. Starting with the third purpose, bare skin with lots of veins offer birds a way to cool efficiently.[227] Probably less important for the African Wattled Lapwing, but certainly key elsewhere, some wattles are used for social signalling, with birds able to rapidly flush them red with blood, similar to humans blushes. Why they should do this we don't really know,[228] and even though Darwin was very excited by human blushes, we still don't properly understand what they're about either. For the African Wattled Lapwing, sexual selection is probably the primary function of wattles today: the size and general impressiveness of such features might be linked genetically to immune function,[229] making wattle size and brightness an honest signal of mate quality: sexy flesh! So for the African Wattled Lapwing caruncles (fleshy lumps like the red bits on the head of this species) and wattles are multipurpose organs, probably primarily used for temperature regulation and as a signal of good genes. For me, though, they're quality comedic appendages.

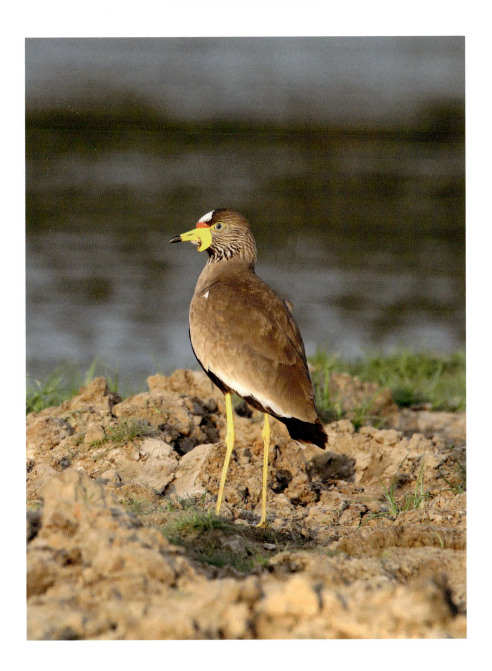

MARABOU STORK
Leptoptilos crumenifer

AIR SACS AND BIRD BREATHING

When it comes to wattles the Marabou Stork looks like it might be a winner, but these pendulous appendages are actually air sacs that this species inflates and deflates at will. This is one of Africa's largest and most remarkable birds. With bare necks, inflatable air sacs front and rear, and a habit of defecating on their legs to cool off, they're not everyone's cup of tea, though! The front air sac is connected directly to the left nostril, the rear to the lungs and so the air sacs are associated with breathing. They can inflate them at will, frequently in courtship, but also as indicators of social dominance: in flocks often only one individual has fully inflated sacs.[230] In fact, all birds have internal air sacs: avian breathing is much more efficient than that of mammals, because a set of air sacs act as bellows and air storage organs, enabling birds to push air in one direction through the lungs – while we have air coming in and out through the same entry point.[231] As alligators have similar air sacs and unidirectional airflow through their lungs, and even fossil dinosaurs show evidence of these features, it seems likely that all dinosaurs had this breathing system. It may have helped them be active when oxygen levels were lower than today.[232] Now it enables birds to breathe efficiently and power active flight even at extreme altitudes, where we would require oxygen. Hence, the external air sacs of Marabou Storks allow us to see the internal air sacs that enable birds to power energetic flight and probably enabled Velociraptors to run so fast and terrorise our mammalian ancestors. I love the way flocks of Marabou appear so rapidly when grass fires begin, and then how they stalk about snapping up toasted grasshoppers while looking somehow like a bunch of old city bankers.

RED-BILLED QUELEA
Quelea quelea

MEGA-FLOCKS

Flocks of Marabou have an uncanny ability to turn up to fires at a moment's notice, but the true flocking specialists are one of Africa's commonest birds: the Red-billed Quelea. With something around 1.5 billion pairs, their total biomass comes to about 60,000 tons, and makes them possibly the most numerous bird in the world. Colonies of two million pairs eat about 40 tons of seeds per day, making this species a significant agricultural pest, although the huge flocks do look spectacular. So why does it have such huge flocks? In grasslands, seeds can become super-abundant as vast areas ripen simultaneously: perfect for supporting massive numbers of quelea. But rain in savannahs is frequently very localised, so ripe seeds are patchily distributed and often only remain on the stems for a few days. Finding these patches as they ripen is crucial: having friends who help search and share information could very helpful indeed. A big daily meeting to exchange intel might be the perfect strategy. In fact, quelea flocks are believed to be one of the best examples of such 'information centre' theories.[233] Behaviour and calls of birds at roosts can broadcast information about their foraging success during the day and other birds can follow successful foragers when the roost disperses, so roosts become communication hubs. Sharing information is only really sensible if your friends return the favour, which is easier if you can identify them: this may be why male Red-billed Quelea have such variable plumage,[234] making it easier to identify reliable informants among the crowd. Naturally, such huge flocks also mean there are many eyes looking out for predators, and if predators do attack the chances of them targeting one specific bird is greatly diluted – two further reasons why many species opt to flock together.

GREAT EGRET
Ardea alba

COMMONNESS AS A RARITY

While the Red-billed Quelea is common and massively abundant, the Great Egret is common by having a huge global distribution. This elegant bird is one of the world's most widespread species, found in the Americas, Europe, Africa, Asia and Australia. While this makes it a common bird, paradoxically such commonness is extremely rare: most species have much lower populations than quelea and smaller ranges than Great Egrets. With a global range of around 368,000,000 km², Great Egrets are probably in the top ten most widely distributed of all bird species.[235] Just 1% of birds have ranges over ten million square kilometres and 50% have ranges that are smaller than the area of France.[236] In fact, 25% of all bird species have ranges that are smaller than the UK. For ornithologists used to northern species, this is remarkable: most European birds are common across the continent, while species with small ranges are crammed into Africa and the rest of the tropics. In Tanzania, the average range size for the birds you can see at any point are about one tenth the area of a similar sample in most of Europe, and in Tanzanian forests the average range size is even smaller. Ultimately, these patterns are generated by processes of speciation, extinction and dispersal, but exactly why there should be so many small-range birds in the tropics and so few species with generally huge ranges is a mystery. My guess is that the long-term environmental stability of much of the tropics is key to explaining this pattern, but there is still a lot of work to do before we have any real evidence.

MADAGASCAR POND HERON
Ardeola idae

TRUE RARITY

At the opposite end of rarity for herons from the Great Egret, the Madagascar Pond Heron is one of East Africa's most endangered species. These are non-breeding visitors to East Africa and have a declining population in Madagascar of around 1,500 pairs, making the species globally Endangered. The Madagascar Pond Heron is rare now, and has probably never been common, but there are several other ways in which species can be rare and endangered. The likes of Lesser Kestrels *Falco naumanni* were once common, but have declined greatly in recent years. Others, such as the Udzungwa Forest-partridge *Xenoperdix udzungwensis* of Tanzania, have populations in the thousands today, but exist in tiny geographical ranges, restricted to just a single mountaintop. Understanding why a once-common species is in decline is often easy (though reversing the process can be far from simple), but explaining why some species are intrinsically rare and others common is hard. For the Madagascar Pond Heron it might be chance: it only breeds in Madagascar, where there are few suitable pools. Without people adding extra pressure on the wetlands in Tanzania and Madagascar, this would probably have been fine – a relatively small population could keep going indefinitely. But for such intrinsically rare species, a little additional pressure can rapidly cause problems. In fact, just as having a large distribution is rare, so is having a large population, and more species are rare than are common: rarity itself is common! Happily, we often know what is required to prevent most extinctions – we just need to find the political will to act.

WHITE-BACKED VULTURE
Gyps africanus

DECLINES, DISEASE CONTROL

Of all the species once common that are now endangered, it is the White-backed Vulture that emblematises the challenges of conservation. This Critically Endangered bird illustrates many issues facing African savannahs: to thrive, they need large, connected landscapes with good mammal populations where they're safe from poisoning. Such landscapes are declining, leading to the critical status of this vulture,[237] though it is happily still frequently seen in Tanzania's protected areas. The movements of this species are truly astonishing: while most individuals travel only small distances, some head off on intrepid journeys across the continent. As vultures follow each other to carcasses, they are extremely vulnerable to poisoning events: a single poisoned carcass can kill hundreds of birds that may have moved great distances. While they may not be the prettiest, White-backed Vultures play an essential role in nutrient cycling: they eat 14,000 tons of meat in Serengeti each year, more than all the mammalian carnivores combined! They probably also play an important role in prevention of diseases, from rabies to anthrax (though the evidence is currently weaker than often presented), so it might seem that their conservation is sensible out of human self-interest alone. In reality, however, if we want to protect vultures, we need ways to maintain and improve savannah landscapes across the entire continent – where millions of people rely on savannahs for their livelihoods. It is not for me as a white European to determine how local people should use their land: conservation is already littered with abuses, as local people are removed in order to facilitate conservation.[238] But I do think finding ways to help communities realise tangible benefits from wildlife that outweigh the very real costs of living with dangerous creatures is an obligation for conservationists.

WOODLAND KINGFISHER
Halcyon senegalensis

WHY DO CONSERVATION?

If vultures show us that self-interest should be one impetus to conserve birds, the Woodland Kingfisher probably offers a more profound reason. Like most of the world's kingfishers, the Woodland Kingfisher is not actually a fisher at all, but eats insects and small land animals, far from water. Although it would be possible to argue that by eating unwanted insect pests human self-interest should again justify conservation, or if not that then such a beautiful bird should be able to draw tourists to come and see it, actually I am unashamedly of the opinion that, ultimately, conservation is a moral obligation. Whether you believe all life has intrinsic value or that value comes from reflecting the works of a creator God doesn't matter to me: I believe the world is better with Woodland Kingfishers and their kind. I think many are like me in this respect, but conservation has developed a rather disingenuous tendency to justify itself solely based on arguments of economic value. Certainly, biodiversity often has significant monetary worth, but that isn't my primary motivation. Financial value for nature comes directly from ecotourism, or indirectly from the costs we'd have to pay to replace the role of an ecosystem in, for example, purifying water, or cleaning the air. But this is a risky basis for conservation: what if artificial water purification gets cheaper? Maybe planted forestry could be made more efficient at clearing air than natural woodlands? Taking a moral view has its own problems, of course, as it somehow implies that the Woodland Kingfisher is global heritage, belonging to no-one or everyone alike. Yet the conservation of this species directly impinges on people who may not share my view, or who simply need firewood – the local communities in the areas where the Woodland Kingfisher lives should surely have greater ownership. This is why conservation often turns to financial value: practically speaking, if Woodland Kingfishers are worth more alive than dead, they may stand a better chance of surviving. Is the question, then, how many Woodland Kingfishers do we need? Or who really owns biodiversity? Either way, it's a stunning bird to watch and since we started the book with a Lilac-breasted Roller prompting thoughts about colourful birds in the tropics, this brings us nicely full circle.

NOTES

1. Cooney, C. R. *et al.* (2022) Latitudinal gradients in avian colourfulness. *Nature Ecology & Evolution* 6, 622–629.
2. Smith, B. T., Seeholzer, G. F., Harvey, M. G., Cuervo, A. M. & Brumfield, R. T. (2017) A latitudinal phylogeographic diversity gradient in birds. *PLoS Biology* 15, e2001073.
3. Fjeldså, J., Bowie, R. C. K. & Rahbek, C. (2012) The role of mountain ranges in the diversification of birds. *Annual Review of Ecology, Evolution & Systematics* 43, 249–265.
4. Day, J. J., Martins, F. C. & Tobias, J. A. (2020) Contrasting trajectories of morphological diversification on continents and islands in the afrotropical white-eye radiation. *Journal of Biogeography* 47, 2235–2247.
5. Gillespie, M. J. *et al.* (2011) Histological and global gene expression analysis of the "lactating" pigeon crop. *BMC Genomics* 12, 452.
6. Gillespie, M. J. *et al.* (2012) Functional similarities between pigeon "milk" and mammalian milk: induction of immune gene expression and modification of the microbiota. *PloS One* 7, e48363.
7. XC83787 Crested francolin (*Dendroperdix sephaena*). https://www.xeno-canto.org/83787.
8. van Niekerk, J. H. (2018) Coalition formation, mate selection and pairing behaviour of the Crested Francolin. *The Ostrich* 89, 71–78.
9. Macleod, R. (2006) Why does diurnal mass change not appear to affect the flight performance of alarmed birds? *Animal Behaviour* 71, 523–530.
10. Thomas, R. J. (1999) Two tests of a stochastic dynamic programming model of daily singing routines in birds. *Animal Behaviour* 57, 277–284.
11. XC469101 Red-and-yellow Barbet (*Trachyphonus erythrocephalus*). https://www.xeno-canto.org/469101.
12. Payne, R. B. & Skinner, N. J. (2008) Temporal patterns of duetting in African barbets. *The Ibis* 112, 173–183.
13. Peach, W. J., Hanmer, D. B. & Oatley, T. B. (2001) Do southern African songbirds live longer than their European counterparts? *Oikos* 93, 235–249.
14. Karr, J. R., Nichols, J. D., Klimkiewicz, M. K. & Brawn, J. D. (1990) Survival rates of birds of tropical and temperate forests: will the dogma survive? *The American Naturalist* 136, 277–291.
15. Lloyd, P., Abadi, F., Altwegg, R. & Martin, T. E. (2014) South temperate birds have higher apparent adult survival than tropical birds in Africa. *Journal of Avian Biology* 45, 493–500.
16. Moreau, R. E. (2008) Clutch-size: a comparative study, with special reference to African birds. *The Ibis* 86, 286–347.
17. Jetz, W., Sekercioglu, C. H. & Böhning-Gaese, K. (2008) The worldwide variation in avian clutch size across species and space. *PLoS Biology* 6, 2650–2657.
18. Hegner, R. E. & Emlen, S. T. (2010) Territorial organization of the White Fronted Bee-eater in Kenya. *Ethology: Formerly Zeitschrift Fur Tierpsychologie* 76, 189–222.
19. Emlen, S. T. & Wrege, P. H. (2010) Forced copulations and intra-specific parasitism: two costs of social living in the white-fronted bee-eater. *Ethology: Formerly Zeitschrift Fur Tierpsychologie* 71, 2–29.
20. Papageorgiou, D. *et al.* (2019) The multilevel society of a small-brained bird. *Current Biology* 29, R1120–R1121.
21. Grueter, C. C. *et al.* (2020) Multilevel organisation of animal sociality. *Trends in Ecology & Evolution* 35, 834–847.
22. Cantor, M. *et al.* (2021) The importance of individual-to-society feedbacks in animal ecology and evolution. *The Journal of Animal Ecology* 90, 27–44.
23. Carstens, K. F., Kassanjee, R., Little, R. M., Ryan, P. G. & Hockey, P. A. R. (2019) Natal dispersal in the Southern Ground Hornbill *Bucorvus leadbeateri*. *The Ostrich* 90, 119–127.
24. Mabry, K. E., Shelley, E. L., Davis, K. E., Blumstein, D. T. & Van Vuren, D. H. (2013) Social mating system and sex-biased dispersal in mammals and birds: a phylogenetic analysis. *PloS One* 8, e57980.

25. Stephen Dobson, F. (1982) Competition for mates and predominant juvenile male dispersal in mammals. *Animal Behaviour* 30, 1183–1192.

26. Carlson, A. (1986) Group territoriality in the Rattling Cisticola, *Cisticola chiniana*. *Oikos* 47, 181–189.

27. Correa, S. M., Adkins-Regan, E. & Johnson, P. A. (2005) High progesterone during avian meiosis biases sex ratios toward females. *Biology Letters* 1, 215–218.

28. Rubenstein, D. R. (2016) Superb Starlings: cooperation and conflict in an unpredictable environment. in *Cooperative Breeding in Vertebrates* (ed. Koenig, W. D., & Dickinson, J.) (Publisher: Cambridge University Press), 181–196.

29. Kuiper, S. M. & Cherry, M. I. (2002) Brood parasitism and egg matching in the Red-chested Cuckoo *Cuculus solitarius* in southern Africa. *The Ibis* 144, 632–639.

30. Cherry, M. I. & Bennett, A. T. (2001) Egg colour matching in an African Cuckoo, as revealed by ultraviolet-visible reflectance spectrophotometry. *Proceedings. Biological Sciences / The Royal Society* 268, 565–571.

31. Honza, M., Kuiper, S. M. & Cherry, M. I. (2005) Behaviour of African turdid hosts towards experimental parasitism with artificial Red-chested Cuckoo *Cuculus solitarius* eggs. *Journal Of Avian Biology* 36, 517–522.

32. Savalli, U. M. (1995) Morphology, territoriality and mating system of the Pin-tailed Whydah *Vidua macroura*. *The Ostrich* 66, 129–134.

33. Barnard, P. E. (1989) Comparative mating system and reproductive ecology of the African whydahs (*Vidua*). (Publisher: University of the Witwatersrand).

34. Spottiswoode, C. N., Begg, K. S. & Begg, C. M. (2016) Reciprocal signaling in honeyguide-human mutualism. *Science* 353, 387–389.

35. Wood, B. M., Pontzer, H., Raichlen, D. A. & Marlowe, F. W. (2014) Mutualism and manipulation in hadza–honeyguide interactions. *Evolution and Human Behavior* 35, 540–546.

36. Bezuidenhout, J. D. & Stutterheim, C. J. (1980) A critical evaluation of the role played by the Red-billed Oxpecker *Buphagus erythrorhynchus* in the biological control of ticks. *The Onderstepoort Journal of Veterinary Research* 47, 51–75.

37. Mooring, M. S., McKenzie, A. A. & Hart, B. L. (1996) Role of sex and breeding status in grooming and total tick load of impala. *Behavioral Ecology and Sociobiology* 39, 259–266.

38. Palmer, M. S. & Packer, C. (2018) Giraffe bed and breakfast: camera traps reveal Tanzanian Yellow-billed Oxpeckers roosting on their large mammalian hosts. *African Journal of Ecology* 56, 882–884.

39. Bishop, A. L. & Bishop, R. P. (2014) Resistance of wild African ungulates to foraging by Red-billed Oxpeckers (*Buphagus erythrorhynchus*): evidence that this behaviour modulates a potentially parasitic interaction. *African Journal of Ecology* 52, 103–110.

40. Nunn, C. L., Ezenwa, V. O., Arnold, C. & Koenig, W. D. (2011) Mutualism or parasitism? using a phylogenetic approach to characterize the oxpecker-ungulate relationship. *Evolution* 65, 1297–1304.

41. Flower, T. (2011) Fork-tailed Drongos use deceptive mimicked alarm calls to steal food. *Proceedings. Biological Sciences / The Royal Society* 278, 1548–1555.

42. Flower, T. P., Gribble, M. & Ridley, A. R. (2014) Deception by flexible alarm mimicry in an African bird. *Science* 344, 513–516.

43. Oatley, T. B. (1969) The functions of vocal imitation by African *Cossyphas*. *The Ostrich* 40, 85–89.

44. Ferguson, J. W. H., van Zyl, A. & Delport, K. (2002) Vocal mimicry in African *Cossypha* robin chats. *Journal Fur Ornithologie* 143, 319–330.

45. Raihani, N. J. & Ridley, A. R. (2008) Experimental evidence for teaching in wild Pied Babblers. *Animal Behaviour* 75, 3–11.

46. Engesser, S., Ridley, A. R. & Townsend, S. W. (2017) Element repetition rates encode functionally distinct information in Pied Babbler "clucks" and "purrs." *Animal Cognition* 20, 953–960.

47. Engesser, S. (2017) Vocal combinations in the Southern Pied Babbler (*Turdoides bicolor*) and the Chestnut-crowned Babbler (*Pomatostomus ruficeps*): implications for the evolution of human language. (PhD Thesis: University of Zurich).

48. Roelofs, Y. (2010) Tool use in birds - an overview of reported cases, ontogeny and underlying cognitive abilities. (MSc Thesis: University of Groningen).

49. Emery, N. J. & Clayton, N. S. (2009) Tool use and physical cognition in birds and mammals. *Current Opinion in Neurobiology* 19, 27–33.

50. Pepperberg, I. M. (1990) Some cognitive capacities of an African Grey Parrot (*Psittacus erithacus*). in *Advances in the Study of Behavior* (eds. Slater, P. J. B., Rosenblatt, J. S. & Beer, C.) (Publisher: Academic Press), vol, 19 357–409.

51. Oldest parrot ever. https://guinnessworldrecords.com/world-records/442525-oldest-parrot-ever.

52. Lopes, A. R. S. *et al.* (2017) The influence of anti-predator training, personality and sex in the behavior, dispersion and survival rates of translocated captive-raised parrots. *Global Ecology and Conservation* 11, 146–157.

53. Wirthlin, M. *et al.* (2018) Parrot genomes and the evolution of heightened longevity and cognition. *Current Biology* 28, 4001-4008.e7.

54. Prassack, K. A., Pante, M. C., Njau, J. K. & de la Torre, I. (2018) The paleoecology of Pleistocene birds from middle bed ii, at Olduvai Gorge, Tanzania, and the environmental context of the Oldowan-Acheulean transition. *Journal of Human Evolution* 120, 32–47.

55. Ducatez, S., Clavel, J. & Lefebvre, L. (2015) Ecological generalism and behavioural innovation in birds: technical intelligence or the simple incorporation of new foods? *The Journal of Animal Ecology* 84, 79–89.

56. Hackett, P. M. W. (2020) Complex avian behaviour and cognition: a mapping sentence approach. in *The Complexity of Bird Behaviour: A Facet Theory Approach* (ed. Hackett, P. M. W.) (Publisher: Springer International Publishing). 67–85.

57. Gottschalk, T. K. (2007) New and notable records of birds from Serengeti national park. *Scopus* 26, 10–21.

58. Beale, C. M., Baker, N. E., Brewer, M. J. & Lennon, J. J. (2013) Protected area networks and savannah bird biodiversity in the face of climate change and land degradation. *Ecology Letters* 16, 1061–1068.

59. Hartley, S. E. & Beale, C. (2019) Impacts of climate change on trophic interactions in grasslands. *Grasslands and Climate Change* 12, 188.

60. Neate-Clegg, M. H. C., Stuart, S. N., Mtui, D., Şekercioğlu, Ç. H. & Newmark, W. D. (2021) Afrotropical montane birds experience upslope shifts and range contractions along a fragmented elevational gradient in response to global warming. *PloS One* 16, e0248712.

61. Neate-Clegg, M. H. C., Stanley, T. R., Şekercioğlu, Ç. H. & Newmark, W. D. (2021) Temperature-associated decreases in demographic rates of afrotropical bird species over 30 years. *Global Change Biology* 27, 2254–2268.

62. deMenocal, P. B. (1995) Plio-pleistocene African climate. *Science* 270, 53–59.

63. Şekercioğlu, Ç. H., Primack, R. B. & Wormworth, J. (2012) The effects of climate change on tropical birds. *Biological Conservation* 148, 1–18.

64. Collias, N. E. & Collias, E. C. (1962) An experimental study of the mechanisms of nest building in a weaverbird. *The Auk* 79, 568–595.

65. Collias, E. C. & Collias, N. E. (1964) The development of nest-building behavior in a weaverbird. *The Auk* 81, 42–52.

66. Howman, H. R. G. & Begg, G. W. (1983) Nest building and nest destruction by the Masked Weaver, *Ploceus velatus. South African Journal Of Zoology* 18, 37–44.

67. Winterbottom, M., Burke, T. & Birkhead, T. R. (1999) A stimulatory phalloid organ in a weaver bird. *Nature* 399, 28–28.

68. Goymann, W., Safari, I., Muck, C. & Schwabl, I. (2016) Sex roles, parental care and offspring growth in two contrasting coucal species. *Royal Society Open Science* 3, 160463.

69. XC426328 White-browed Coucal (*Centropus superciliosus*). https://www.xeno-canto.org/426328.

70. Goymann, W. (2020) Males paving the road to polyandry? parental compensation in a monogamous nesting cuckoo and a classical polyandrous relative. *Ethology: Formerly Zeitschrift Fur Tierpsychologie* 126, 436–444.

71. Komeda, S. (1983) Nest attendance of parent birds in the Painted Snipe (*Rostratula benghalensis*). *The Auk* 100, 48–55.

72. Liker, A., Freckleton, R. P. & Székely, T. (2013) The evolution of sex roles in birds is related to adult sex ratio. *Nature Communications* 4, 1587.

73. Andersson, S. (1989) Sexual selection and cues for female choice in leks of Jackson's Widowbird *Euplectes jacksoni. Behavioral Ecology and Sociobiology* 25, 403–410.

74. Galván, I., Camarero, P. R., Mateo, R. & Negro, J. J. (2016) Porphyrins produce uniquely ephemeral animal colouration: a possible signal of virginity. *Scientific Reports* 6, 39210.

75. Lichtenberg, E. M. & Hallager, S. (2008) A description of commonly observed behaviors for the Kori Bustard (*Ardeotis kori*). *Journal of Ethology* 26, 17–34.

76. Virani, M. Z. & Harper, D. M. (2009) Factors influencing the breeding performance of the Augur Buzzard *Buteo augur* in southern Lake Naivasha, Rift Valley, Kenya. *The Ostrich* 80, 9–17.

77. Koivula, M. & Korpimäki, E. (2001) Do scent marks increase predation risk of microtine rodents? *Oikos* 95, 275–281.

78. Lind, O., Mitkus, M., Olsson, P. & Kelber, A. (2013) Ultraviolet sensitivity and colour vision in raptor foraging. *The Journal of Experimental Biology* 216, 1819–1826.

79. Jones, M. P., Pierce, K. E. & Ward, D. (2007) Avian vision: a review of form and function with special consideration to birds of prey. *Journal of Exotic Pet Medicine* 16, 69–87.

80. Karaman, S. & Frazzoli, E. (2012) High-speed flight in an ergodic forest. in *2012 IEEE International Conference on Robotics and Automation* (Publisher: Institute of Electrical and Electronics Engineers). 2899–2906.

81. Potier, S. *et al.* (2017) Eye size, fovea, and foraging ecology in accipitriform raptors. *Brain, Behavior and Evolution* 90, 232–242.

82. Jackson, A. L., Ruxton, G. D. & Houston, D. C. (2008) The effect of social facilitation on foraging success in vultures: a modelling study. *Biology Letters* 4, 311–313.

83. Jackson, C. R. *et al.* (2020) A dead giveaway: foraging vultures and other avian scavengers respond to auditory cues. *Ecology and Evolution* 10, 6769–6774.

84. Kane, A. & Kendall, C. J. (2017) Understanding how mammalian scavengers use information from avian scavengers: cue from above. *The Journal of Animal Ecology* 86, 837–846.

85. Payne, R. S. (1971) Acoustic location of prey by Barn Owls (*Tyto alba*). *The Journal of Experimental Biology* 54, 535–573.

86. Krumm, B., Klump, G., Köppl, C. & Langemann, U. (2017) Barn Owls have ageless ears. *Proceedings. Biological Sciences / The Royal Society* 284, 20171584.

87. Thewissen, J. G. M. & Nummela, S. (2008) *Sensory evolution on the threshold: adaptations in secondarily aquatic vertebrates* (Publisher: University of California Press).

88. Casler, C. L. (1973) The air-sac systems and buoyancy of the Anhinga and Double-crested Cormorant. *The Auk* 90, 324–340.

89. Cohn, J. W. (1969) Adaptations for locomotion and feeding in the Anhinga and the Double-crested Cormorant. *The Auk* 86, 154–155.

90. Hennemann, W. W. (1988) Energetics and spread-winged behavior in Anhingas and Double-crested Cormorants: the risks of generalization. *Integrative and Comparative Biology* 28, 845–851.

91. Stenhagen, E. & Odham, G. (1971) Chemistry of preen gland waxes of waterfowl. *Accounts of Chemical Research* 4, 121–128.

92. Rijke, A. M. & Jesser, W. A. (2011) The water penetration and repellency of feathers revisited. *The Condor* 113, 245–254.

93. Liu, Y., Chen, X. & Xin, J. H. (2008) Hydrophobic duck feathers and their simulation on textile substrates for water repellent treatment. *Bioinspiration & Biomimetics* 3, 046007.

94. Stolen, P. & McKinney, F. (1983) Bigamous behaviour of captive Cape Teal. *Wildfowl* 34, 10–13.

95. Brennan, P. L. R. *et al.* (2007) Coevolution of male and female genital morphology in waterfowl. *PloS One* 2, e418.

96. Anthony, S. J. *et al.* (2013) A strategy to estimate unknown viral diversity in mammals. *MBio* 4, e00598-13.

97. Chan, J. F.-W., To, K. K.-W., Tse, H., Jin, D.-Y. & Yuen, K.-Y. (2013) Interspecies transmission and emergence of novel viruses: lessons from bats and birds. *Trends in Microbiology* 21, 544–555.

98. Gaidet, N. *et al.* (2008) Evidence of infection by H5N2 highly pathogenic avian influenza viruses in healthy wild waterfowl. *PLoS Pathogens* 4, e1000127.

99. Simmons, R. E., Barnard, P. & Jamieson, I. G. (1999) What precipitates influxes of wetland birds to ephemeral pans in arid landscapes? observations from Namibia. *The Ostrich* 70, 145–148.

100. Roshier, D. A., Klomp, N. I. & Asmus, M. (2006) Movements of a nomadic waterfowl, Grey Teal *Anas gracilis*, across inland Australia - results from satellite telemetry spanning fifteen months. *Ardea* 94, 461–475.

101. Bettinetti, R. *et al.* (2011) A preliminary evaluation of the DDT contamination of sediments in Lakes Natron and Bogoria (eastern Rift Valley, Africa). *Ambio* 40, 341–350.

102. Krienitz, L. *et al.* (2003) Contribution of hot spring cyanobacteria to the mysterious deaths of lesser flamingos at Lake Bogoria, Kenya. *FEMS Microbiology Ecology* 43, 141–148.

103. Ballot, A. *et al.* (2004) Cyanobacteria and cyanobacterial toxins in three alkaline Rift Valley lakes of Kenya—Lakes Bogoria, Nakuru and Elmenteita. *Journal of Plankton Research* 26, 925–935.

104. Childress, B. *et al.* (2004) Satellite tracking lesser flamingo movements in the Rift Valley, East Africa: pilot study report. *The Ostrich* 75, 57–65.

105. Bakaloudis, D. E. & Vlachos, C. G. (2011) Feeding habits and provisioning rate of breeding Short-toed Eagles *Circaetus gallicus* in northeastern Greece. *Journal of Biological Research* 16, 166–176.

106. Lerner, H. R. L. & Mindell, D. P. (2005) Phylogeny of eagles, old world vultures, and other accipitridae based on nuclear and mitochondrial DNA. *Molecular Phylogenetics and Evolution* 37, 327–346.

107. Darawshi, S., Motro, U. & Leshem, Y. (2009) *The ecology of the Short-toed Eagle (Circaetus gallicus) in the Judean Slopes Israel.* Report to Rufford Foundation.

108. Fry, C. H. (2008) The recognition and treatment of venomous and non-venomous insects by small bee-eaters. *The Ibis* 111, 23–29.

109. Karageorgi, M. *et al.* (2019) Genome editing retraces the evolution of toxin resistance in the Monarch butterfly. *Nature* 574, 409–412.

110. Rothschild, M. & Kellett, D. N. (2009) Reactions of various predators to insects storing heart poisons (cardiac glycosides) in their tissues. *Journal of Entomology* 46, 103–110.

111. Lewis, D. C., Metallinos-Katzaras, E. & Grivetti, L. E. (1987) Coturnism: human poisoning by European migratory quail. *Journal of Cultural Geography* 7, 51–65.

112. Kennedy, B. W. & Grivetti, L. E. (1980) Toxic quail: a cultural-ecological investigation of coturnism. *Ecology of Food and Nutrition* 9, 15–41.

113. Sokolovskis, K. *et al.* (2018) Ten grams and 13,000 km on the wing - route choice in Willow Warblers *Phylloscopus trochilus yakutensis* migrating from far East Russia to East Africa. *Movement Ecology* 6, 20.

114. Jensen, F. P., Falk, K. & Petersen, B. S. (2006) Migration routes and staging areas of Abdim's Storks *Ciconia abdimii* identified by satellite telemetry. *The Ostrich* 77, 210–219.

115. Webster, M. S., Marra, P. P., Haig, S. M., Bensch, S. & Holmes, R. T. (2002) Links between worlds: unraveling migratory connectivity. *Trends in Ecology & Evolution* 17, 76–83.

116. Burgess, N. D. & Mlingwa, C. O. F. (2000) Evidence for altitudinal migration of forest birds between montane eastern arc and lowland forests in East Africa. *The Ostrich* 71, 184–190.

117. Pearson, D., Backhurst, G. & Jackson, C. (2014) The study and ringing of palaearctic birds at Ngulia lodge, Tsavo West national park, Kenya, 1969–2012: an overview and update. *Scopus: Journal of East African.*

118. Nwaogu, C. J. & Cresswell, W. (2016) Body reserves in intra-African migrants. *Journal of Ornithology* 157, 125–135.

119. Brooke, R. K. & Herroelen, P. (1988) The nonbreeding range of southern African bred European Bee-eaters *Merops apiaster. The Ostrich* 59, 63–66.

120. Ramos, R. *et al.* (2016) Population genetic structure and long-distance dispersal of a recently expanding migratory bird. *Molecular Phylogenetics And Evolution* 99, 194–203.

121. Åkesson, S. *et al.* (2017) Timing avian long-distance migration: from internal clock mechanisms to global flights. *Philosophical Transactions of The Royal Society of London. Series B, Biological Sciences* 372, 20160252.

122. Cassone, V. M. & Westneat, D. F. (2012) The bird of time: cognition and the avian biological clock. *Frontiers in Molecular Neuroscience* 5, 32.

123. Thorup, K. & Holland, R. A. (2009) The bird GPS - long-range navigation in migrants. *The Journal of Experimental Biology* 212, 3597–3604.

124. Gwinner, E. & Scheuerlein, A. (1999) Photoperiodic responsiveness of equatorial and temperate-zone stonechats. *The Condor* 101, 347–359.

125. Dittami, J. P. & Gwinner, E. (2009) Annual cycles in the African Stonechat *Saxicola torquata axillaris* and their relationship to environmental factors. *Journal of Zoology* 207, 357–370.

126. Dawson, A., King, V. M., Bentley, G. E. & Ball, G. F. (2001) Photoperiodic control of seasonality in birds. *Journal of Biological Rhythms* 16, 365–380.

127. Ferrero, J. J., Grande, J. M. & Negro, J. J. (2003) Copulation behavior of a potentially double-brooded bird of prey, the Black-winged Kite (*Elanus caeruleuas*). *The Journal of Raptor Research* 37, 1–7.

128. Byrom, A. E. *et al.* (2015) Small mammal diversity and population dynamics in the greater Serengeti ecosystem. in *Serengeti IV - Sustaining Biodiversity in a Coupled Human–Natural System* (eds. Sinclair, A. R. E., Metzger, K. L., Fryxell, J. M. & Mduma, S. A. R.) (Publisher: University of Chicago Press), 323–358.

129. Negro, J. J. *et al.* (2006) Convergent evolution of *Elanus* kites and the owls. *Journal of Raptor Research* 40, 222–225.

130. Pijanowski, B. C. (1992) A revision of lack's brood reduction hypothesis. *The American Naturalist* 139, 1270–1292.

131. Simmons, R. (2010) Offspring quality and the evolution of cainism. *The Ibis* 130, 339–357.

132. Margalida, A., García, D., Heredia, R. & Bertran, J. (2010) Video-monitoring helps to optimize the rescue of second-hatched chicks in the endangered Bearded Vulture *Gypaetus barbatus. Bird Conservation International* 20, 55–61.

133. Simon Thomsett on the African Crowned Eagle – part 3. http://www.africanraptors.org/simon-thomsett-on-the-african-crowned-eagle-part-3/.

134. Shultz, S. & Noë, R. (2002) The consequences of Crowned Eagle central-place foraging on predation risk in monkeys. *Proceedings. Biological Sciences / The Royal Society* 269, 1797–1802.

135. Berger, L. R. & Clarke, R. J. (1995) Eagle involvement in accumulation of the Taung child fauna. *Journal of Human Evolution* 29, 275–299.

136. Boshoff, Palmer & Avery. (1990) Variation in the diet of Martial Eagles in the Cape Province, South Africa. *South African Journal of Wildlife Research* 20, 57–68.

137. Roemer, G. W., Gompper, M. E. & Van Valkenburgh, B. (2009) The ecological role of the mammalian mesocarnivore. *Bioscience* 59, 165–173.

138. Ritchie, E. G. & Johnson, C. N. (2009) Predator interactions, mesopredator release and biodiversity conservation. *Ecology Letters* 12, 982–998.

139. Seavy, N. E. & Apodac, C. K. Raptor abundance and habitat use in a highly-disturbed-forest landscape in western Uganda. *J. Raptor Res.* 36, 51–57.

140. Buij, R., Croes, B. M., Gort, G. & Komdeur, J. (2013) The role of breeding range, diet, mobility and body size in associations of raptor communities and land-use in a West African savanna. *Biological Conservation* 166, 231–246.

141. Soobramoney, S., Downs, C. T. & Adams, N. J. (2004) Variability in foraging behaviour and prey of the Common Fiscal Shrike, *Lanius collaris*, along an altitudinal gradient in South Africa. *The Ostrich* 75, 133–140.

142. Classen, A. *et al.* (2014) Complementary ecosystem services provided by pest predators and pollinators increase quantity and quality of coffee yields. *Proceedings. Biological Sciences / The Royal Society* 281, 20133148.

143. Howe, H. F. (1990) Seed dispersal by birds and mammals. in *Reproductive Ecology of Tropical Forest Plants* (eds. Bawa, K. S. & Hadley, M.) (Publisher: UNESCO and The Parthenon Publishing Group). 191–218.

144. Moreau, R. E. (1958) Some aspects of the musophagidae. *The Ibis* 100, 238–270.

145. Sharam, G. J., Sinclair, A. R. E. & Turkington, R. (2009) Serengeti birds maintain forests by inhibiting seed predators. *Science* 325, 51.

146. Buregyeya, H. *et al.* (2014) Role of birds and bats in long distance transmission of banana bacterial wilt in Uganda. *International Journal of Agriculture Innovations and Research* 2, 636–640.

147. Newmark, W. D., Mkongewa, V. J., Amundsen, D. L. & Welch, C. (2020) African sunbirds predominantly pollinate plants useful to humans. *The Condor* 122, duz070.

148. Ollerton, J. (1998) Sunbird surprise for syndromes. *Nature* 394, 726–727.

149. Rodríguez-Gironés, M. A. & Santamaría, L. (2004) Why are so many bird flowers red? *PLoS Biology* 2, e350.

150. Niknafs, S. & Roura, E. (2018) Nutrient sensing, taste and feed intake in avian species. *Nutrition Research Reviews* 31, 256–266.

151. Kare, M. (1971) Comparative study of taste. in *Taste* (eds. Acree, T. E. et al.) (Publisher: Springer Berlin Heidelberg), 278–292.

152. Purchase, C., Nicolson, S. W. & Fleming, P. A. (2013) Salt intake and regulation in two passerine nectar drinkers: White-bellied Sunbirds and New Holland Honeyeaters. *Journal of Comparative Physiology. B, Biochemical, Systemic, and Environmental Physiology* 183, 501–510.

153. Tewksbury, J. J. & Nabhan, G. P. (2001) Directed deterrence by capsaicin in chilies. *Nature* 412, 403–404.

154. Thompson, M. (2015) How birds make colorful feathers. https://academy.allaboutbirds.org/how-birds-make-colorful-feathers/ (2015).

155. Hu, D. *et al.* (2018) A bony-crested Jurassic dinosaur with evidence of iridescent plumage highlights complexity in early paravian evolution. *Nature Communications* 9, 217.

156. Withgott, J. (2000) Taking a bird's-eye view…in the UV: recent studies reveal a surprising new picture of how birds see the world. *Bioscience* 50, 854–859.

157. Bennettand, P. M. & Harvey, P. H. (1987) Active and resting metabolism in birds: allometry, phylogeny and ecology. *Journal of Zoology* 213, 327–344.

158. Brooks, D. M. (2012) Birds caught in spider webs: a synthesis of patterns. *The Wilson Journal of Ornithology* 124, 345–353.

159. Sibly, R. M. *et al.* (2012) Energetics, lifestyle, and reproduction in birds. *Proceedings of the National Academy of Sciences of the United States of America* 109, 10937–10941.

160. McKechnie, A. E., Körtner, G. & Lovegrove, B. G. (2006) Thermoregulation under semi-natural conditions in Speckled Mousebirds: the role of communal roosting. *African Zoology* 41, 155–163.

161. Smit, B. & McKechnie, A. E. (2010) Do owls use torpor? Winter thermoregulation in free-ranging Pearl-spotted Owlets and African Scops-owls. *Physiological and Biochemical Zoology: PBZ* 83, 149–156.

162. Stevens, M., Troscianko, J., Wilson-Aggarwal, J. K. & Spottiswoode, C. N. (2017) Improvement of individual camouflage through background choice in ground-nesting birds. *Nature Ecology & Evolution* 1, 1325–1333.

163. Troscianko, J., Wilson-Aggarwal, J., Stevens, M. & Spottiswoode, C. N. (2016) Camouflage predicts survival in ground-nesting birds. *Scientific Reports* 6, 19966.

164. Cott, H. B. & Benson, C. W. (1969) The palatability of birds, mainly based upon observations of a tasting panel in Zambia. *The Ostrich* 40, 357–384.

165. Götmark, F. (1994) Are bright birds distasteful? a re-analysis of H. B. Cott's data on the edibility of birds. *Journal of Avian Biology* 25, 184–197.

166. Brown, L. H. & Urban, E. K. (2008) The breeding biology of the Great White Pelican *Pelecanus onocrotalus roseus* at Lake Shala, Ethiopia. *The Ibis* 111, 199–237.

167. Cushman, G. T. (2013) *Guano and the Opening of the Pacific World: A Global Ecological History*. (Publisher: Cambridge University Press).

168. Eden, T. E. (1846) *The Search for Nitre, and the True Nature of Guano: Being an Account of a Voyage to the South-west Coast of Africa : Also a Description of the Minerals Found There, and of the Guano Islands in that Part of the World*. (Publisher: R. Groombridge).

169. Beolens, B. & Watkins, M. (2003) *Whose Bird? Men and Women Commemorated in the Common Names of Birds*. (Publisher: Helm).

170. Stoner, J. L., Loken, B. & Stadler Blank, A. (2018) The name game: how naming products increases psychological ownership and subsequent consumer evaluations. *Journal of Consumer Psychology* 28, 130–137.

171. Smith, C. D. (1994) The roots of Rwandan genocide. *Refuge: Canada's Journal on Refugees* 14, 13–14.

172. Heneberg, P. *et al.* (2016) Conservation of the Red Kite *Milvus milvus* (Aves: accipitriformes) is not affected by the establishment of a broad hybrid zone with the Black Kite *Milvus migrans migrans* in central Europe. *PloS One* 11, e0159202.

173. Scheider, J., Wink, M., Stubbe, M. & Hille, S. (2004) Phylogeographic relationships of the Black Kite *Milvus migrans*. in *Raptors Worldwide* (eds. Chancellor, R. D. & Meyburg, B.-U.) (Publisher: World Working Group on Birds of Prey and Owls & MME/BirdLife Hungary), 467–472.

174. Donald, P. F. & Collar, N. J. (2011) Notes on the structure and plumage of Beesley's Lark. *Bulletin of the African Bird Club* 18, 168–173.

175. Alström, P. *et al.* (2013) Multilocus phylogeny of the avian family Alaudidae (larks) reveals complex morphological evolution, non-monophyletic genera and hidden species diversity. *Molecular Phylogenetics and Evolution* 69, 1043–1056.

176. Manegold, A. & Podsiadlowski, L. (2014) On the systematic position of the Black-collared Lovebird *Agapornis swindernianus* (Agapornithinae, psittaciformes). *Journal of Ornithology* 155, 581–589.

177. Ottenburghs, J., Ydenberg, R. C., Van Hooft, P., Van Wieren, S. E. & Prins, H. H. T. (2015) The avian hybrids project: gathering the scientific literature on avian hybridization. *The Ibis* 157, 892–894.

178. Mandiwana-Neudani, T. G., Little, R. M., Crowe, T. M. & Bowie, R. C. K. (2019) Taxonomy, phylogeny and biogeography of African spurfowls galliformes, phasianidae, phasianinae, coturnicini: *Pternistis* spp. *The Ostrich* 90, 145–172.

179. Price, T. D. & Bouvier, M. M. (2002) The evolution of F1 postzygotic incompatibilities in birds. *Evolution; International Journal of Organic Evolution* 56, 2083–2089.

180. Taylor, S. A. & Larson, E. L. (2019) Insights from genomes into the evolutionary importance and prevalence of hybridization in nature. *Nature Ecology & Evolution* 3, 170–177.

181. Sankararaman, S., Mallick, S., Patterson, N. & Reich, D. (2016) The combined landscape of Denisovan and Neanderthal ancestry in present-day humans. *Current Biology* 26, 1241–1247.

182. Oliveros, C. H. *et al.* (2019) Earth history and the passerine superradiation. *Proceedings Of The National Academy of Sciences of the United States of America* 116, 7916–7925.

183. Jønsson, K. A. *et al.* (2010) Biogeographical history of cuckoo-shrikes (Aves: passeriformes): transoceanic colonization of Africa from Australo-papua. *Journal of Biogeography* 37, 1767–1781.

184. Patten, M. A. & Lasley, G. W. (2000) Range expansion of the Glossy Ibis in North America. *North American Birds* 54, 241–247.

185. Haddrath, O. & Baker, A. J. (2001) Complete mitochondrial DNA geonome sequences of extinct birds: ratite phylogenetics and the vicariance biogeography hypothesis. *Proceedings of the Royal Society of London. Series B: Biological Sciences* 268, 939–945.

186. Prum, R. O. *et al.* (2015) A comprehensive phylogeny of birds (*Aves*) using targeted next-generation DNA sequencing. *Nature* 526, 569–573.

187. Alexander, R. M., Maloiy, G. M. O., Njau, R. & Jayes, A. S. (2009) Mechanics of running of the Ostrich (*Struthio camelus*). *Journal of Zoology* 187, 169–178.

188. Brennan, P. L. R. & Prum, R. O. (2012) The erection mechanism of the ratite penis. *Journal of Zoology* 286, 140–144.

189. Bertram, B. C. R. (1992) *The Ostrich Communal Nesting System*. (Publisher: Princeton University Press).

190. Boire, D., Dufour, J. S., Théoret, H. & Ptito, M. (2001) Quantitative analysis of the retinal ganglion cell layer in the Ostrich, *Struthio camelus*. *Brain, Behavior And Evolution* 58, 343–355.

191. Burt, D. B., Coulter, P. F. & Ligon, D. (2007) Evolution of parental care and cooperative breeding. in *Reproductive Biology and Phylogeny of Birds, Part B: Sexual Selection, Behavior, Conservation, Embryology and Genetics* (ed. Jamieson, B. G. M.) (Publisher: CRC Press). 305–336.

192. Botelho, J. F. & Faunes, M. (2015) The evolution of developmental modes in the new avian phylogenetic tree. *Evolution & Development* 17, 221–223.

193. Ar, A. & Yom-Tov, Y. (1978) The evolution of parental care in birds. *Evolution* 32, 655–669.

194. Probert, J. R. *et al.* (2019) Anthropogenic modifications to fire regimes in the wider Serengeti-mara ecosystem. *Global Change Biology* 25, 3406–3423.

195. See note 163 for a photograph of Temminck's Courser eggs.

196. Sandgathe, D. M., & Berna, F. (2017). Fire and the genus *Homo*: An introduction to supplement 16. *Current Anthropology* 58, S165-S174.

197. Thomas, D. H. & Robin, A. P. (2009) Comparative studies of thermoregulatory and osmoregulatory behaviour and physiology of five species of sandgrouse (Aves: pterocliidae) in morocco. *Journal of Zoology* 183, 229–249.

198. MacLean, G. L. (1983) Water transport by sandgrouse. *Bioscience* 33, 365–369.

199. Cade, T. J. & Maclean, G. L. (1967) Transport of water by adult sandgrouse to their young. *The Condor* 69, 323–343.

200. See Note 197.

201. Tattersall, G. J., Andrade, D. V. & Abe, A. S. (2009) Heat exchange from the toucan bill reveals a controllable vascular thermal radiator. *Science* 325, 468–470.

202. van de Ven, T. M. F. N., Martin, R. O., Vink, T. J. F., McKechnie, A. E. & Cunningham, S. J. (2016) Regulation of heat exchange across the hornbill beak: functional similarities with toucans? *PloS One* 11, e0154768.

203. Stanback, M., Richardson, D. S., Boix-Hinzen, C. & Mendelsohn, J. (2002) Genetic monogamy in Monteiro's Hornbill, *Tockus monteiri*. *Animal Behaviour* 63, 787–793.

204. Mills, M. S. L., Boix-Hinzen, C. & Plessis, M. A. D. U. (2004) Live or let live: life-history decisions of the breeding female Monteiro's Hornbill *Tockus monteiri*. *The Ibis* 147, 48–56.

205. Wilson, R. T. & Wilson, M. P. (1986) Nest building by the Hamerkop *Scopus umbretta*. *The Ostrich* 57, 224–232.

206. Kahl, M. P. (2008) Observations on the behaviour of the Hamerkop *Scopus umbretta* in Uganda. *The Ibis* 109, 25–32.

207. Wilson, R. T., Wilson, M. P. & Durkin, J. W. (2008) Aspects of the reproductive ecology of the Hamerkop *Scopus umbretta* in central Mali. *The Ibis* 129, 382–388.

208. Liversidge, R. (1963) The nesting of the Hamerkop *Scopus umbretta*. *The Ostrich* 34, 55–62.

209. See Note 205

210. Cockle, K. L., Martin, K. & Wesołowski, T. (2011) Woodpeckers, decay, and the future of cavity-nesting vertebrate communities worldwide. *Frontiers in Ecology and the Environment* 9, 377–382.

211. Farah, G., Siwek, D. & Cummings, P. (2018) Tau accumulations in the brains of woodpeckers. *PloS One* 13, e0191526.

212. Wang, L. *et al.* (2011) Why do woodpeckers resist head impact injury: a biomechanical investigation. *PloS One* 6, e26490.

213. Swennen, C. & Yu, Y.-T. (2004) Notes on feeding structures of the Black-faced Spoonbill *Platalea minor*. *Ornithological Science* 3, 119–124.

214. Kopij, G. (1997) Breeding ecology of the African Spoonbill *Platalea alba* in the Free State, South Africa. *The Ostrich* 68, 77–79.
215. Fodor, A. (1970) The origins of the Arabic legends of the pyramids. *Acta Orientalia Academiae Scientiarum Hungaricae* 23, 335–363.
216. Ikram, S. (2015) *Divine creatures: animal mummies in ancient Egypt* (Publisher: the American University in Cairo press).
217. Curtis, C., Millar, C. D. & Lambert, D. M. (2018) The Sacred Ibis debate: the first test of evolution. *PLoS Biology* 16, e2005558.
218. Tate, G. J., Bishop, J. M. & Amar, A. (2016) Differential foraging success across a light level spectrum explains the maintenance and spatial structure of colour morphs in a polymorphic bird. *Ecology Letters* 19, 679–686.
219. Shawkey, M. D. & Hill, G. E. (2004) Feathers at a fine scale. *The Auk* 121, 652–655.
220. Karell, P., Ahola, K., Karstinen, T., Valkama, J. & Brommer, J. E. (2011) Climate change drives microevolution in a wild bird. *Nature Communications* 2, 208.
221. Portugal, S. J., Murn, C. P., Sparkes, E. L. & Daley, M. A. (2016) The fast and forceful kicking strike of the Secretary Bird. *Current Biology* 26, R58–R59.
222. Blanco, R. E. & Jones, W. W. (2005) Terror birds on the run: a mechanical model to estimate its maximum running speed. *Proceedings. Biological Sciences / The Royal Society* 272, 1769–1773.
223. Kahl, M. P. (1971) Food and feeding behavior of Openbill storks. *Journal Fur Ornithologie* 112, 21–35.
224. Gosner, K. L. (1993) Scopate tomia: an adaptation for handling hard-shelled prey? *The Wilson Bulletin* 105, 316–324.
225. Rand, A. L. (1954) On the spurs on birds' wings. *The Wilson Bulletin* 66, 127–134.
226. Berglund, A. (2013) Why are sexually selected weapons almost absent in females? *Current Zoology* 59, 564–568.
227. Nolan, W. F., Weathers, W. W. & Sturkie, P. D. (1978) Thermally induced peripheral blood flow changes in chickens. *Journal of Applied Physiology: Respiratory, Environmental and Exercise Physiology* 44, 81–84.
228. Negro, J. J., Sarasola, J. H., Fariñas, F. & Zorrilla, I. (2006) Function and occurrence of facial flushing in birds. *Comparative Biochemistry and Physiology. Part A, Molecular & Integrative Physiology* 143, 78–84.
229. Baratti, M., Ammannati, M., Magnelli, C., Massolo, A. & Dessì-Fulgheri, F. (2010) Are large wattles related to particular MHC genotypes in the male pheasant? *Genetica* 138, 657–665.
230. Akester, A. R., Pomeroy, D. E. & Purton, M. D. (2009) Subcutaneous air pouches in the Marabou Stork (*Leptoptios curmeniferus*). *Journal of Zoology* 170, 493–499.
231. Duncker, H. R. (1974) Structure of the avian respiratory tract. *Respiration Physiology* 22, 1–19.
232. Brocklehurst, R. J., Schachner, E. R. & Sellers, W. I. (2018) Vertebral morphometrics and lung structure in non-avian dinosaurs. *Royal Society Open Science* 5, 180983.
233. Bayer, R. D. (1982) How important are bird colonies as information centers? *The Auk* 99, 31–40.
234. Dale, J. (2000) Ornamental plumage does not signal male quality in Red-billed Queleas. *Proceedings. Biological Sciences / The Royal Society* 267, 2143–2149.
235. BirdLife International. (2021) Species factsheet: *Ardea alba*. http://datazone.birdlife.org/species/factsheet/great-white-egret-ardea-alba (2021).
236. Orme, C. D. L. *et al.* (2006) Global patterns of geographic range size in birds. *PLoS Biology* 4, e208.
237. Ogada, D. *et al.* (2016) Another continental vulture crisis: Africa's vultures collapsing toward extinction. *Conservation Letters* 9, 89–97.
238. Brockington, D., Igoe, J. & Schmidt-Soltau, K. (2006) Conservation, human rights, and poverty reduction. *Conservation Biology* 20, 250–252.

PHOTOGRAPHY CREDITS

Acknowledgements are given by page *Special thanks to Per Holmen* (Holmen-birding-safari.com) *for supplying so many of the images that are core to this book.*

Alan W. Wells 146
Alexander Rees 153, 159
A.E. Hopkins 45, 156
Becky Matsubara CC-BY-2.0 155
Bernard Dupont 27, 35, 61, 114, 121
Bradley R. Hacker 16, 100, 171
Bruce Ward-Smith 72, 158
Chad Rosenthal CC-BY-2.0 193
Chris Sayers 37
Christoph Moning 98
Claude Melde 13, 51, 57, 154, 163, 168,
    170, 172, 173, 188, 191, 197
Colin Beale viii, 1, 10, 12, 20, 32, 34, 46,
    80, 86, 89, 110, 125, 128, 142, 145, 166,
    189, 194
Daniel J. Field 197
Daudi Peterson 9, 69, 79
David Williams 65
Derek Keats 22, 42, 64, 81, 85, 87, 91, 93,
    101, 108, 112, 119, 150, 151, 161, 167,
    178, 187, 190
Francesco Veronesi 47
Gerry Capelin 97, 107
Jacques Gravelin 181
Jean-Louis Carlo 148
Johan van Rensburg 99, 195
Larry Hubble 19
Lorenzo Barelli 49
Michael Pierce 83
Mike Buckham 55, 73, 135
Natasha Peters 74, 200
Nick Athanas 131
Nik Borrow 36
Per Holmen 3, 4, 5, 7, 8, 14, 15, 17, 18, 21,
    23, 24, 28, 29, 31, 33, 39, 40, 43, 44, 48,
    50, 52, 54, 56, 58, 59, 62, 75, 77, 78, 82,
    84, 88, 92, 94, 96, 103, 104, 105, 106,
    109, 111, 113, 116, 122, 123, 124, 127,
    129, 130, 134, 136, 139, 149, 152, 165,
    169, 175, 176, 179, 183, 185, 186, 192,
    198, 201, 203
Pete Steward 25, 66, 95, 132, 140, 164, 177,
    199
Raymond Birkelund 6

Ron Eggert 11, 60, 137
Sadik Kassam 53
Shane McPherson 67
Takashi Muramatsu, flickr 157
Teddy Kinyanjui 38
Tom Conzemius 2, 26, 30, 41, 63, 68, 70,
    71, 76, 90, 102, 115, 117, 118, 120, 126,
    138, 141, 143, 144, 160, 162, 174, 182,
    184, 196
Victoria Beale 133

INDEX

References to photographs appear in *italic*.
Names of the 101 featured species are given in **bold.**

Also available from Pelagic

Low-Carbon Birding, edited by Javier Caletrío

The Hen Harrier's Year, Ian Carter and Dan Powell

Birds & Flowers: An Intimate 50 Million Year Relationship (coming spring 2024), Jeff Ollerton

Reflections: What Wildlife Needs and How to Provide It, Mark Avery (coming autumn 2023)

Traffication: How Cars Destroy Nature and What We Can Do About It, Paul F. Donald

Reconnection: Fixing Our Broken Relationship with Nature, Miles Richardson

Treated Like Animals: Improving the Lives of the Creatures We Own, Eat and Use, Alick Simmons

Invisible Friends: How Microbes Shape Our Lives and the World Around Us, Jake M. Robinson

Wildlife Photography Fieldcraft: How to Find and Photograph UK Wildlife, Susan Young

Rhythms of Nature: Wildlife and Wild Places Between the Moors, Ian Carter

Ancient Woods, Trees and Forests: Ecology, Conservation and Management, edited by Alper H. Çolak, Simay Kırca and Ian D. Rotherham

Essex Rock: Geology Beneath the Landscape, Ian Mercer and Ros Mercer

Wild Mull: A Natural History of the Island and its People, Stephen Littlewood and Martin Jones

Challenges in Estuarine and Coastal Science, edited by John Humphreys and Sally Little

A Natural History of Insects in 100 Limericks, Richard A. Jones and Calvin Ure-Jones

pelagicpublishing.com